T0190476

Environmental Footprints and Eco-design of Products and Processes

Series editor

Subramanian Senthilkannan Muthu, SGS Hong Kong Limited,
Hong Kong, Hong Kong SAR

More information about this series at http://www.springer.com/series/13340

Subramanian Senthilkannan Muthu
Miguel Angel Gardetti

Editors

Green Fashion

Volume 1

Editors
Subramanian Senthilkannan Muthu
Environmental Services Manager-Asia
SGS Hong Kong Limited
Hong Kong
Hong Kong SAR

Miguel Angel Gardetti
Center for Studies on Sustainable Luxury
Buenos Aires
Argentina

ISSN 2345-7651   ISSN 2345-766X   (electronic)
Environmental Footprints and Eco-design of Products and Processes
ISBN 978-981-10-9083-7   ISBN 978-981-10-0111-6   (eBook)
DOI 10.1007/978-981-10-0111-6

Printed on acid-free paper

Springer Science+Business Media Singapore Pte Ltd. is part of Springer Science+Business Media
(www.springer.com)

Preface

Sustainable or Green fashion (also called Eco fashion) is one of the most happening trends in the field of sustainability. The fashion industry is facing a severe threat due to its vulnerable environmental impacts and ruthless social impacts. One of the ways to address this threat is to inculcate "Green" or "Sustainable" or "Eco" practices in the entire fashion industry supply chain and in the entire life cycle of fashion products. Green fashion enforces environmental and social responsibility considerations from the creation and production processes of a fashion product and ensures the same till the entire life cycle of the product under question. This is again a very vast topic and includes a huge list of elements. The agenda of Green fashion is pretty big and we try to cover it in two volumes. This is the first volume of the Green Fashion book, which aims to cover a good range of topics with the aid of nine lively and well-written chapters.

This book disseminates the knowledge on Green Fashion to the readers via nine informative chapters in Volume 1: *Cubreme® and Sustainable Value Creation: A Diagnosis; Facets of Indigo: Combining Traditional Dye Methods with State-of-the-Art Digital Print Technology, A Sustainable Design Case, Understanding Consumer Behavior in the Sustainable Clothing Market: Model Development and Verification; The Feasibility of Large-Scale Composting of Waste Wool; Sustainable Value Generation Through Post-Retail Initiatives: An Exploratory Study of Slow and Fast Fashion Businesses; Hanji, the Mulberry Paper Yarn Rejuvenates Nature and the Sustainable Fashion Industry of Korea; Sustainable Production Processes in Textile Dyeing; and Developments in Sustainable Chemical Processing of Textiles.*

We are highly confident that readers of the book will get a lot of very useful information pertaining to Green fashion from this contribution. We would like to record our sincere thanks to all the authors who have contributed the nine chapters in this book for their time and priceless efforts spent.

Contents

Cubreme® and Sustainable Value Creation: A Diagnosis

Miguel Angel Gardetti

Abstract Cubreme® is a small textile company created by designer Alejandra Gottelli whose purpose is to promote the use of organic and natural textiles by designing classical contemporary garments that transcend fashion trends. Therefore, the collections are not divided into 6-week "seasons" that change continuously, but rather they are featured in line with cold and warm seasons. The brand was conceived in response to the designer's need to express creatively the culture of Argentine native communities. The fibers used by Alejandra Gottelli come from the shearing of domestic species, such as animals from the sheep, camel, and goat families from the Andean–Patagonian and Andean–Cuyo regions of Argentina, which are bred in their natural habitat. The fibers obtained from shearing animals from both the camel and sheep families, as well as the harvest of vegetable fibers are treated in premium spinning mills that develop highly refined products on a very small scale. Fabrics are developed using handlooms, and craft tailor shops are in charge of the final tailoring to give garments a *haute couture* finish; this helps keep a small production line, using renewable resources and contributing both actively and voluntarily to social, economic, and environmental improvement. This case study introduces Cubreme and then shows the model of sustainable value creation that integrates four elements: environment, innovation, stakeholder management, economic value and potential of growth (Hart, Harvard Bus Rev 75: 66–76, 1997; Capitalism at the Crossroads, 2005; Capitalism at the crossroads: capitalism at the crossroads—aligning business, earth, and humanity,

---

The author would like to thank Mrs. Ana Laura Torres for her contributions to this chapter.

M.A. Gardetti (✉)
Center for Studies on Sustainable Luxury, Av. San Isidro 4166,
PB "A", C1429ADP, Buenos Aires, Republic of Argentina
e-mail: mag@lujosustentable.org
URL: http://www.lujosustentable.org

S.S. Muthu and M.A. Gardetti (eds.), *Green Fashion*,
Environmental Footprints and Eco-design of Products and Processes,
DOI 10.1007/978-981-10-0111-6_1

2007; Hart and Milstein, MIT Sloan Manag Rev 41:23–33,1999, Acad Manag Exec 17: 56–67, 2003) as a tool to diagnose the brand, ending with an analysis of the company in the light of the above model and a few conclusions.

Keywords Fashion · Sustainability · Cubreme® · Argentina

1 Introduction

In terms of environmental pollution and social impact, fashion is one of the most harmful systems (Fletcher 2008, 2014; Kozlowsky et al. 2012; Muthu 2014). Over the past years, these negative effects have been heightened by a phenomenon known as "fast fashion,"[1] which is having an impact even at a cultural level. In this regard, this phenomenon is sometimes called "McFashion,"[2] given its total homogenization resulting from the sector's globalization (Lee 2003).

Factors such as scale expansion and increased time to market, in addition to the division of labor, and consumption patterns, have deeply influenced how fashion is produced (Hawley 2011).

The fashion production system, basically driven by economic factors, is causing an intensive use of resources[3] and large volumes of textile wastes with a great impact on the planet's capacity to regenerate (McCann 2015). Retailers and multinational brands have embraced practices based on trend-oriented design, and a lower production cost policy that has taken the massive search for supplies to developing countries. Therefore, this significantly reduces both clothes pricing and quality (Cataldi et al. 2010).

From the consumer's perspective, the continuous cycle of purchase, use, and disposal of clothes also has serious consequences for both society and the environment[4] (Gwilt and Rissanen 2011). Due to their eagerness to follow the latest trends and attracted by affordable prices, consumers' demand is driving the purchase of clothes beyond their actual needs, which results in overconsumption.

---

[1]According to Cataldi et al. (2010), this term refers to the clothing industry focused on low-cost mass production where seasons change every 6 weeks instead of following the two traditional annual seasons. It is sold by retailers at very low prices and based on the latest trends, which encourage consumers to purchase more than they really need, thus resulting in both social and environmental impacts.

[2]The term *McFashion*—the textile equivalent of fast food—was coined making an allusion to the fast food restaurant chain to refer to this fact whereby it is possible to find the same garment in any of the major cities in the world.

[3]An emblematic example is the drought of the Aral Sea due to the indiscriminate and inefficient use of water for cotton crops, among others (Allwood et al. 2006; Fletcher 2008).

[4]As an example, a research study conducted by Cambridge University reveals that, on average, English consumers send 30 kg of garments and textiles per capita to landfills every year (Allwood et al. 2006).

Likewise, consumers have lost their ability to claim their own sense of style, following instead the mandates of brands and trendsetters (Hawley 2011). According to Cataldi et al. (2010), consumers have taken a leading role in the development of the current fast fashion system due to their attitude and behavior in line with production speed.

As described thus far, fashion has become the very antithesis of sustainability (Fletcher and Tham 2015). Fashion needs to escape the fate of being a tool that often encourages excessive economic growth to the detriment of the environmental and social impact. In Fletcher's own words (2008), fashion "should help cultivate new aspirations" (p. 118).

This case analyzes the company under the model of sustainable value creation developed by Professor Stuart L. Hart (Hart 1997; Hart and Milstein 1999, 2003; Hart 2005, 2007) that integrates four aspects: environment, innovation, stakeholder management, and potential for growth.

2 Methodology

In order to develop this case, the author first made a bibliographic compilation on this topic and then researched information about Cubreme® on two levels: through trade media and, on the other hand, through semistructured interviews with designers.

3 Understanding "Sustainable Fashion"

The term "sustainable development" dates back to the United Nations Conference on the Human Environment in 1972 where it was first coined. Sustainable development is a problematic expression on the meaning of which few people agree. Each person can take the term and "reinvent" it considering his or her own needs. This is a concept that continuously leads us to change objectives and priorities because it is an open process and as such, it cannot be reached definitely. However, one of the most widely accepted definitions of sustainable development, although diffuse and nonoperating, is the one proposed by the World Commission on Environment and Development (WCED) report, *Our Common Future*, which defines sustainable development as "the development model that allows us to meet the present needs, without compromising the ability of future generations to meet their own needs" (WCED 1987: 43). At its core is the notion that all natural systems have limits and that human well-being requires living within those limits. The essential objective of this development model is to raise the quality of life with the long-term maximization of the productive potential of ecosystems with the appropriate and relevant technologies (Gardetti 2005).

According to this report the three pillars of sustainability would be "people, profit and planet" (Bader 2008). Sustainable development is not only a new concept, but also a new paradigm, and this requires looking at things in a different way. It is a notion of the world deeply different from the one that dominates our current thinking and includes satisfying basic human needs such as justice, freedom, and dignity (Ehrenfeld 1999).

Although the term "fashion" refers to products such as clothing and accessories, according to Fletcher (2008) and (2014) fashion is the way in which our clothes reflect and communicate our individual vision within society, linking us to time and space. Clothing is the material thing that gives fashion a contextual vision in society (Cataldi et al. 2010). According to Hethorn and Ulasewicz (2008), fashion is a process that is expressed and worn by people, and as a material object, has a direct link to the environment. It is embedded in everyday life. Therefore, sustainability within fashion means that the development and use of some thing or process is not harmful to people or the planet, and once put into action, such thing or process can enhance the well-being of those people who interact with it, and the environment within which it is developed and used. But this is not always the case. One of the most worrying social effects in the current fashion system is that to meet the market's expectations in terms of turnaround and competitive prices the conditions in which clothes are made fail to comply with minimum labor standards: respect for workers, fair salaries, working hours, breaks, and health and safety standards in the work environment (Oxfam GB, Clean Clothes Campaign & ICFTU 2004). Clothing manufacture has moved to countries that pay lower salaries and where there are neither stringent pollution controls nor laws that punish textile companies that cause it (Hethorn and Ulasewicz 2008).

The textile and fashion industries use large quantities of water and energy (two of the resources of major concern worldwide), and also generate waste, effluents, and pollution. Both manufacture and consumption of textile products are significant sources of environmental damage.[5]

Furthermore, another environmental implication of the current fashion system is associated with transport. More than in any other industry, a textile or fashion product is made up of different components that come from every corner of the world. The fiber/yarn is produced in one country, then shipped to be spun in another country, then shipped again to separate production and finishing processes, before finally being made up into garments somewhere else (Earley 2007). This certainly increases CO_2 emission rates, which ultimately have a negative impact on climate change (Muthu 2014; Farrer 2011).

Fashion at its worst promotes materialism, because marketing and advertising techniques have helped it be perceived as a means to achieve both success and happiness. It is also involved in very serious health conditions, such as bulimia and anorexia (Fletcher 2008, 2014). Moreover, the pressure of constantly reshaping

---

[5]Other authors and organizations have studied and analyzed the textile and clothing industry environmental impacts too. Some of them are Slater (2000), UK Department for Environment, Food and Rural Affairs (2008), Ross (2009), Dickson et al. (2009) and Tobler-Rohr (2011).

our personal identity, incited by changing trends, creates both personal insecurity and stress. To us, the meaning of fashion may be to purchase things that we do not really need, use them very little, and quickly dispose of them. A vast proportion of clothes are now bought in supermarkets, and new collections arrive in High Street stores every 6–8 weeks. Although purchasing new products other than what we really need supports a production system based on economic growth, it has a negative impact on the resource base and undermines personal satisfaction. The ongoing shaping and reshaping of our identity through fashion consumption is both a distinctive feature of the times we live in and a key cause of the sector's lack of sustainability (Earley 2007).

The strategy implemented in the design of contemporary manufactured goods—aesthetic obsolescence[6]—ensures that producers continue producing and consumers continue buying (Fuad-Luke 2004–2005).

Some authors, such as Walker (2006) and Koefoed and Skov (year unknown), have studied the contradictions between fashion and sustainability: sustainability requires a drastic reduction in our ecological footprint, and increasing a product's useful life. Fashion, on the other hand, suggests a passing trend or fad: something transient, superficial, and often rather wasteful. But, beyond these contradictions, fashion should not necessarily come into conflict with sustainable principles. Indeed, fashion plays a role in the promotion and achievement of sustainability, and it may even be a key to more sustainable ways of living. Thus sustainable fashion is an approach to the fashion system intended to minimize negative environmental impacts, and, in turn, maximize positive impacts (benefits) for workers and their families all along the value chain, hence playing a decisive role in poverty reduction. For this reason, Kate Fletcher (2012) in the preface of the book, *Sustainability in Fashion and Textiles: Values, Design, Production and Consumption*, wrote, "For me the fostering of alternatives to the status quo in fashion and textiles is essential if we are to deeply engage with the process of sustainability…" (p. ix).

4 Creating Sustainable Value

The challenges associated with global sustainability, viewed through the appropriate set of business lenses, can help identify strategies and practices that promote the creation of value. Michael Porter and Mark R. Kramer in their work "Creating Share Value" of 2011 say, "[T]he purpose of the corporation must be redefined as creating shared value, not just profit per se…It will also reshape capitalism and its relationship to society." That is, there is a connection between the social dimension and economic growth (Porter and Kramer 2006; Fatemi

---

[6]Term that Vance Packard made popular in *The Waste Makers* (1963) to refer to manufacturers' strategy to render products outdated, nonfunctional, or useless after a period of time, estimated in advance during the design phase. The purpose of planned obsolescence is to get quick economic profit.

and Fooladi undated; Gholami 2011), and this is demonstrated in the work, "Sustainability and Competitive Advantage: An Empirical Study of Value Creation," conducted in 2011 by Gupta and Benson on American companies. Value creation needs to be broadly defined to acknowledge the strategically relevant stakeholders of a corporation, and the definition of value creation is an interactive process that includes stakeholders (Freeman 1984; Freeman et al. 2007; Post et al. 2002; Sachs et al. 2008; Sachs and Maurer 2009). For all this, the *sustainable enterprise* represents the potential for a new approach in bringing the private sector closer to development, including poverty, the respect for cultural diversity, and the preservation of ecological integrity (Hart 2005, 2007).

4.1 The Creation of Value

With this model, both short- and long-term value is created using two variables: a spatial variable and a temporal one.

The temporal variable reflects the firm's need to manage "today's" business, while simultaneously creating "tomorrow's" technologies and markets. In contrast, the spatial variable reflects the firm's need to nurture and protect "internal" organizational skills, technologies, and capabilities, while simultaneously providing the firm with new prospects and knowledge from "outside" stakeholders (Hart 1997, 2005, 2007; Hart and Milstein 2003).

The combination of these two variables (see Fig. 1) results in four different dimensions, crucial to the creation of value (Hart and Milstein 2003):

- *Internal* dimension and *immediate* term, such as cost and risk reduction
- *External* dimension and *immediate* term (building of legitimacy)
- *Future* dimension (or *long* term) and *internal* (innovation and repositioning)
- *Long-term* dimension and *external* (credible expectations of growth)

Fig. 1 Dimensions for value creation. *Source* Designed by the authors (Adapted from Stuart L. Hart and Mark Milstein, *Creating Sustainable Value* 2003)

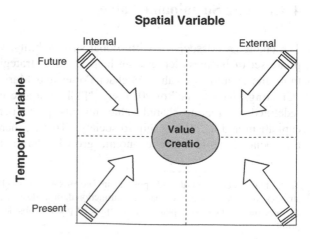

To maximize value creation companies must act efficiently and simultaneously in the four dimensions.

4.2 Global Drivers for Sustainability

According to Hart (1997), (2005), (2007) and Hart and Milstein (2003), there are four groups of drivers related to global sustainability shown in Fig. 2 and explained below.

The first group corresponds to the growth of industrialization and its associated impacts, such as consumption of materials, pollution, and waste and effluent generation. Thus, efficiency in the use of resources and pollution prevention are crucial to sustainable development.

A second group of drivers is associated with the proliferation and interconnection of civil society stakeholders, with high expectations placed on business performance beyond their economic action. To achieve sustainable development, companies are challenged to operate in an open, responsible, and informed manner.

The third group of drivers regarding global sustainability is related to emerging technologies that would provide radical and "disturbing" solutions and that could render many of today's energy- and material-intensive industries obsolete.

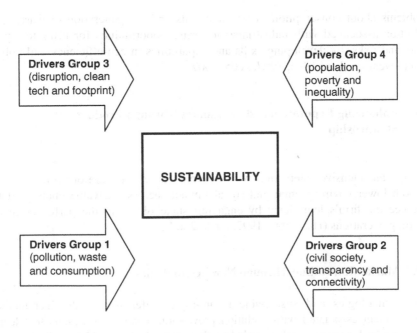

Fig. 2 Sustainability drivers. *Source* Designed by the author

Thus, innovation and technological change are the keys to achieving sustainable development.

Finally, the fourth group of drivers is linked to population growth. In addition, economic globalization affects the local autonomy, the culture, and the environment, causing a growing decline in developing countries (Hart 2005, 2007; Hart and Milstein 2003). A long-term vision that incorporates (the traditional economic aspects) social and environmental aspects is essential for the achievement of sustainable development.

4.3 The Sustainable Value Structure: Connecting Drivers with Strategies

Global sustainability is a complex multidimensional concept that cannot be addressed by any single corporate action. The creation of (sustainable) value implies that firms have to manage the four groups of drivers (Hart 1997, 2005, 2007; Hart and Milstein 2003). Each group of drivers has a strategy and practice, which correspond to a particular dimension of value creation.

4.3.1 Growing Profits and Reducing Risk Through Pollution Prevention

Problems about consumption of raw materials and the generation of waste and pollution associated with industrialization raise opportunities for firms to lower costs and risks, by developing skills and capabilities in eco-efficiency and pollution prevention (Hart 1995, 1997, 2005, 2007).

4.3.2 Enhancing Reputation and Legitimacy Through Product Stewardship

Product stewardship integrates the voice of stakeholders into business processes through an intensive interaction with external parties. It therefore offers a way to both lower environmental and social impacts across the value chain, and to enhance the firm's legitimacy by engaging stakeholders in the performance of ongoing operations (Hart 1995, 1997, 2005, 2007).

4.3.3 Market Innovation Through New Technologies

New technologies and sustainable technologies refer not to the incremental improvement associated with pollution prevention, but to innovations that leapfrog standard routines and knowledge (Hart and Milstein 1999). Thus rather than simply seeking to reduce the negative impacts of their operations, firms can strive to solve social and environmental problems with the internal development or acquisition of new capabilities that address the sustainability challenge directly (Hart 1997, 2005, 2007; Hart and Milstein 2003).

4.3.4 Crystallizing the Growth Path Through the Sustainability Vision

The vision of sustainability, which creates a map for tomorrow's businesses, provides members of the organization with the necessary guidance in terms of organizational priorities, technological development, resource allocation, and design of business models (Hart and Milstein 2003).

This model highlights the nature and magnitude of those possibilities associated with sustainable development and relates them to the creation of value for the company. This appears in Fig. 3 which shows the strategies and practices associated with the creation of both short- and long-term value.

4.4 A Tool for Diagnosis

In order to choose which strategy(s) to promote (see Fig. 4) and analyze the best way to manage it or them, the sustainable value structure may be used as a simple but important diagnosis tool. By evaluating a company's activity in each of the quadrants, the portfolio balance level can be evaluated (Hart 1997, 2005, 2007).

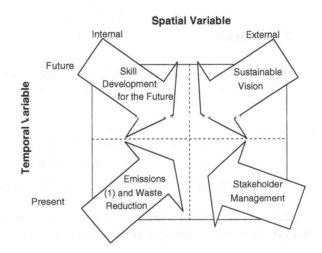

Fig. 3 Strategies and practices associated with the creation of short- and long-term value, and the essential elements for its development. *Source* Designed by the authors (Adapted from Stuart I. Hart and Mark Milstein, "Creating Sustainable Value" 2003)

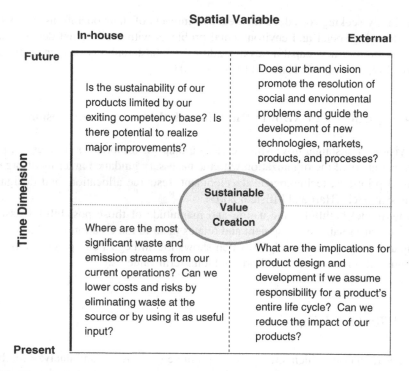

Fig. 4 Evaluation of each of the sustainability portfolio quadrants. *Source* Designed by the author (Adapted from Stuart L. Hart, "Beyond Greening: Strategies for a Sustainable World," 1997; Stuart L. Hart and Mark Milstein, "Creating Sustainable Value," 2003 and Stuart L. Hart, "Capitalism at the Crossroads," 2005 and 2007)

5 Cubreme®

"Sustainable fashion is for the long term. It's a path made by experimentation."
"I'm tenacious when I believe in something, in this case, blending fashion with sustainability."
"I don't follow the crowd."
"I've never liked settling for what's available."
"I am a believer in a change of paradigm. I do my utmost for it."

Alejandra Gotelli, Founder and CEO, Cubreme®

This small textile company (adhered to the United Nations Global Compact[7] and, for some years now, certified with B[8] System) was created by designer Alejandra Gottelli whose purpose is to promote the use of organic and natural textiles by designing classical-contemporary garments that transcend fashion trends.

---

[7]www.unglobalcompact.org.

[8]http://www.sistemab.org/ingles/home.

Therefore, the collections are not divided into 6-week "seasons" that change continuously, but rather they are featured in line with cold and warm seasons.

The brand was conceived in response to the designer's need to express the culture of Argentine native communities creatively. Her sister was a biologist who lived for a long time in Ethiopia and who told her stories about aboriginal communities of that country. After listening to the tales and appreciating the fabrics and decorative pieces that her sister used to bring, Alejandra compared the African handicrafts to those from the north of Argentina, and that stirred up her interest in developing a project to create a 100 % Argentine-made textile product, thus revaluating a sector of the economy hit hard mainly during the 1990s due to the then-prevailing economic situation. That idea was the kick-off of a learning process that took her to different places in the country.

The animal fibers used by Alejandra Gottelli come from the shearing of domestic species, such as animals from the sheep and camel families from the Andean–Patagonian and Andean–Cuyo regions of Argentina, which are bred in their natural habitat. Vegetable fibers such as agro-ecological cotton come from the provinces of the northeastern region of Argentina (Chaco and Corrientes), where small producers grow their crops using a biodiversity approach, protecting the environment, and ensuring the livelihood of farming families and communities.[9] Therefore, Cubreme is a proposal based on agro-ecological principles and an inclusive development model, with both autonomy and equity. Currently, a change is in process towards organic cotton from Peru.

Fabrics are developed using hand looms, and craft tailor shops are in charge of the final tailoring to give garments a *haute couture* finish; all this helps keep a small production line, using renewable resources and contributing both actively and voluntarily to social, economic, and environmental improvement.

Her premise was to create from the raw material at hand. Although she decided to use hand looms, which are traditional in Latin America and have the potential to preserve ancient techniques and practices that promote social inclusion, her goal was to avoid developing telluric products but, instead, items imbued with an "urban, light, and versatile" style.

Little by little, all the links in the chain—raw material producers, spinners, and weavers—became bound together. All these players know one another and the rest of the chain, which largely helps them appreciate both other people's and joint work.

At first, Cubreme was focused on creating coats using loom-woven fabrics, but soon the designer noticed that there was a yarn remainder which could not be woven on the loom due to technical reasons (small volumes). Then, she developed a knit line, employing knitters who were at home and use their semi-industrial knitting machines or who were working part time. In addition, she purchased machines lent for use agreement to some knitters in order to optimize garment

---

[9]Worked with Cooperativa Agroecológica del Litoral Ltda., certified with the Fair Trade Label Organisation (FLO), Native dyeable Cotton.

production and quality, and to cut down working hours using more efficient, functional, and modern knitting machines.

Her value chain consists of the following.

Raw Materials

Renewable raw materials are made up of Argentine natural fibers from agricultural producers associated with cooperatives. Alejandra Gottelli believes that cooperativism is an inclusion model and a tool for the social transformation of small producers and workers. She also uses linen from Belgium and Brazil, bamboo from China, and organically certified llama fiber from La Carolina, in the Province of San Luis (Argentina).

Animal fibers, such as wool from Merino sheep, are sourced from OVIS 21. This company is focused on increasing cattle-raising profitability and regenerating grazing lands. It offers training, consulting, and product certification services. As the hub of the Savory Institute, it specializes in holistic management, in addition to having vast experience in sheep and wool.

Cubreme obtained a permit from the national agency in charge of wild fauna in order to use vicuña and guanaco fibers to develop a luxury clothing line. For such purpose, the company began working with Cooperativa Payun Matrú, which is engaged in the protection and preservation of guanaco populations, as well as in improving the lives of the inhabitants of the Reservation La Payunia, in the province of Mendoza.

Figure 5 shows the shearing of sheep, on the left, and the (Fig. 6) shearing of guanacos, on the right.

Spinning Process

The fibers obtained from shearing animals from both the camel and sheep families, as well as the harvest of vegetable fibers are treated in premium spinning mills, Almafuerte, which develop highly refined products at very small scale, in compliance with the controls and standards of organic certifications (Organización Internacional Agropecuaria, OIA [International Agricultural and Cattle Organisation], and Global Organic Textile Standards, GOTS).

Handcraft Tailoring

In line with the designer's guidelines, fabric creation is in charge of weavers with vast experience thanks to a long track record in hand loom weaving, training, and even loom make. Figure 7 shows a hand loom.

This process is intended to have a new appreciation of local handicrafts, giving each fabric a unique and impeccable finish. For instance, garments are cut one at a time. Before cutting, fabrics are bonded[10] and decatized[11] in order to provide them with structure and softness. Cubreme works with two workshops that combined allocate five people to this task.

---

[10]Finishing process which consists in "bonding" the fabrics to improve their structure.

[11]Process which results in a smooth, wrinkle-free, and soft finish for worsted or woolen yarn fabrics.

Fig. 5 Shearing of sheep (*left*)

Fig. 6 Shearing of guanacos (*right*). *Source* Cubreme®. Published with the designer's authorization

Fig. 7 Hand Loom. *Source*
Cubreme. Published with the
designer's authorization

Final Product

Garment design and tailoring take place at local craft tailor shops (both for men
and women) and spinning mills. Each model is carefully developed relying on
an approach to caution and responsible consumption. The designer is actively
involved in the entire process. Figure 8 shows a tailor at work, and final products
are shown in Figs. 9 and 10.

Over the past few years, Cubreme has developed a home decor line consisting
of llama hair rugs, throws (for bed and couch), aprons, and beach mats, the last
two made of "native" cotton.

Customers

Most customers are tourists with a strong sustainable awareness, both Americans
and Europeans who jointly account for 70 % of revenues. Thirty percent are local
customers, defined as captive by the designer, as they "feel the spirit of the brand."
As to the age group, Alejandra Gotelli explains that it is highly diverse: from
young people to 60-year-olds.

Fig. 8 At the tailor's. *Source* Cubreme. Published with the designer's authorization

Fig. 9 Finished product (woman's coat). *Source* Cubreme. Published with the designer's authorization

Fig. 10 CS overcoat and ghorvat coat. *Source* Cubreme. Published with the designer's authorization

Alejandra Gottelli believes that "[I]t takes too much time and dedication to develop natural fibers, and they are so difficult to get that garments must have durability: clothes are conceived to be worn for 10–20 years." These concepts reflect the values intended to be conveyed through the clothes: the distinction and loyalty of durable items. Thus, at her store, the designer and entrepreneur explains to her customers that "Sustainable fashion shouldn't be a fad: it should be a transformation path. This involves creating a network, trying to find solutions co-created by organizations, cooperatives, companies and designers who share the same philosophy." In her opinion, sharing information is the only way to help sustainable fashion to grow.

From an economic perspective Cubreme has been over its break-even point since 2010, less than a year from when this undertaking was kicked off. Three per cent of sales income is donated to Nutri Red, an organization that acts as a meeting point, an information node for organizations working around the malnutrition issue in Argentina.

6 Creating Sustainable Value in Cubreme: Diagnosis and Conclusions

It is not easy truly to apply the concept of sustainability to the fashion system. Many companies in the fashion industry regard this concept as one more trend and resort to it as a marketing strategy to get quick economic benefits.

Far from this distorted and reductionist approach, the design of truly sustainable fashion means changing this goal and striking a balance between economic profitability and the promotion of environmental and social quality. In the fashion system, these are complex issues and improvements result from a mix of creativity, good judgment, and information about the processes dealt with and the ability to take a life cycle approach to bring about change. Working with environmentally friendly materials is not enough. Designers should acquaint themselves with not only their processes, but also with those of their value chain, developing a holistic and comprehensive approach that identifies interdependences and synergies, and helps find opportunities for whole-system improvement (Fletcher 2008, 2014). Cubreme is taking this path by combining sustainable sourced natural raw materials and promoting cooperativism and fair trade as a way further to improve the quality of life of local communities. Each link in the value chain offers an opportunity to implement environmental and social considerations: from the use of natural fibers as raw materials, through the spinning process in certified spinning mills, and the fabric creation stage that rescues the craft of loom weaving, to the final garment tailoring phase that revalues the tailor's trade. Therefore, the designer manages to achieve a perfect blend of her sensitivity and her social and environmental commitment in the fashion creative process.

It should be noted that, because the brand adheres to both the UNGC and B System, it gains—through joint network-coordinated work within the framework of these two initiatives—new capabilities that help it innovate and have a sustainable vision of the future. Figure 11 shows a brief summary of the strategies and practices currently developed by the brand in order to create sustainable value.

To maximize sustainable value creation Cubreme must act efficiently and simultaneously in the four dimensions so it is worth seeing to what extent this is a

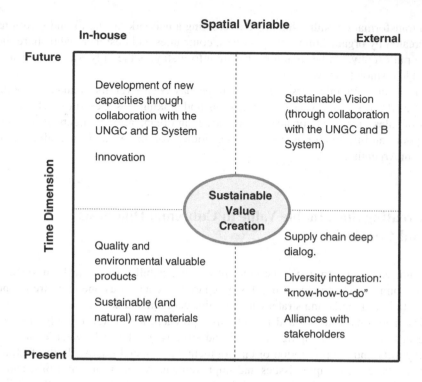

Fig. 11 Strategies and practices for creating sustainable value in Cubreme. *Source* Designed by the author

balanced model. Based on Fig. 4 and the questions on that figure,[12] this evaluation is made by assigning a score to each quadrant and their questions. For example, (1) nonexistent, (2) emerging, (3) set, and (4) institutionalized.[13]

In connection with the lower left-hand quadrant (internal-today), although Cubreme uses natural ("sustainable") raw materials, it should review the use of organic cotton due to the huge water consumption and land use required by this kind of fiber.[13] The brand has no water consumption, power, or process-generated

---

[12]This analysis has a limitation as it only uses the questions contained in Fig. 4 and some degree of bias from the author when each quadrant is evaluated.

[13]If we focus, with a holistic vision (as required by sustainability), on the food crisis that has afflicted humanity for many decades now, which is only looming larger, we should take into account that organic cotton production requires significantly more land (and more water?) than conventional cotton, as well as more work. This obviously results in higher costs, thus becoming a quasi-exclusive product, and restricting its access to a small population segment. This may seem morally questionable if we think that the areas where both types of (conventional and organic) cotton are grown are, generally, regions with high poverty rates and, sometimes abject poverty. Source: Reflexiones sobre el Algodón Orgánico, unpublished document, Gardetti MA and Torres AL, 2012.

emission assessment, mainly in terms of those raw materials sourced from other counties (organic cotton from Peru, linen from Belgium and Brazil, and bamboo from China), or from Argentina though from places far away from Cubreme operations (e.g., wool from Merino sheep from OVIS 21). Based on the above, it may be qualified as an "emerging" (2) quadrant.

As to the lower right-hand quadrant (external-today), the brand has the capacities required to innovate, and thus, to reduce the impact of its products. However, there is an evident flaw when it comes to the entire product life cycle. Cubreme is closely related to both the B System and the UNGC, however, it would be advisable eventually to build relationships with other stakeholders so as to help the brand include greater "diversity" missing today. The stage of this quadrant would be somewhere between "nonexistent and emerging" (1.5).

In terms of the upper left-hand quadrant (internal-tomorrow), the firm has the potential to implement creative improvements, but if we take the life cycle of the product it develops, to a certain extent product sustainability may be limited by the current competence base. It is an "emerging" (2) quadrant.

As to the upper right-hand quadrant (external-tomorrow), the designer is on the right track to make her vision promote social and environmental solutions. As a designer, Alejandra Gotelli is more focused on design, product, and processes. Her ideas could be shared with other designers in order to take her vision to the next level. Moreover, she fails to develop new markets more in line with her products' quality, characteristics, and prices, which would include "sustainable luxury." This quadrant might be located very close to the "emerging" status (approximately 2).

Figure 12 depicts the balance degree of the sustainable value creation model. This balance can be defined as "reasonably balanced," which means that Cubreme is simultaneously performing in the four quadrants[14] in a balanced way. Figure 13 shows the practices and strategies that the brand should integrate into a process, setting priorities to improve its sustainable performance.

As pointed out by Farrer (2011), it might be utopian to think that sustainability is possible in terms of mass market in the current fashion industry, which is highly

---

[14]An unbalanced portfolio (model) is a sign of problems. A model tilted to the lower part suggests the brand is well positioned, but it may be vulnerable in the future. A portfolio tilted to the upper part indicates there is a sustainability vision, but it lacks the operating or analytical abilities for implementation. A model titled to the left quadrant indicates a concern about social and environmental challenge management with improved internal processes and technology development initiatives. Finally, a portfolio tilted to the right runs the risk of being considered socially and environmentally shallow, because the main operations still cause serious environmental damages (Hart 1997, 2005, 2007). An unbalanced portfolio also suggests missed opportunities and vulnerability. There are still few fashion labels that recognize sustainability strategic opportunities. These companies focus on and allocate their time to the lower half of the portfolio, which implies short-term solutions considering the existing products and the different stakeholder groups (Hart 1997, 2005, 2007).

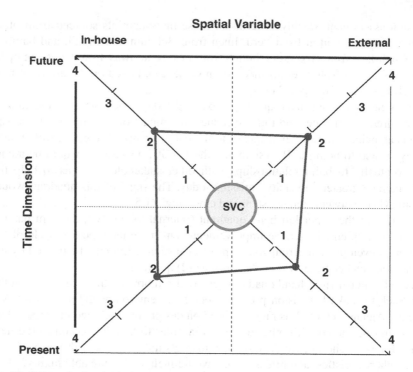

Fig. 12 Diagnosis (model balance) in Cubreme. *Source* Designed by the author. *Note* SVC (Sustainable Value Creation)

globalized and features extremely fragmented production processes. Nevertheless, the stage is set for the onset of a new fashion system in which the message that it is possible, and necessary, to develop both new business models and lifestyles is brewing among producers and users. In such a scheme, designers play a major role in promoting a different production system whereby the industry is made up of small volumes as in the case of Cubreme.

Even today, many companies do not recognize the strategic opportunities that result from sustainability. They are normally focused on, and allocate their time to, short-term solutions, looking at the existing products and the different groups of stakeholders. However, Cubreme is developing sustainable ideas that need to be supplemented by other practices and strategies to create more sustainable value.

Author's Note

Once the development and presentation of this case was completed, People for the Ethical Treatment of Animals (PETA) filed a complaint for animal cruelty against the firm OVIS21 which provides "sustainable" wool to Cubreme. Due to

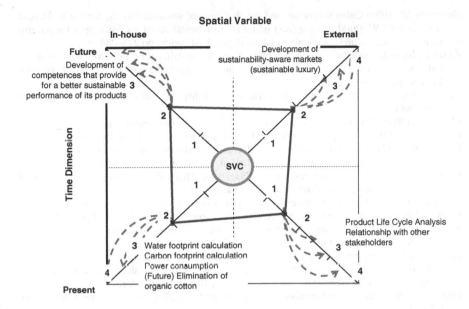

Fig. 13 Practices and strategies missing in Cubreme for improved sustainable performance. SVC (sustainable value creation); *red dotted lines* Cubreme's potential improvement in each quadrant by implementing the strategies and practices shown in the figure. *Source* Designed by the author (color on online)

such complaint, some brands, such as Stella McCartney and Patagonia, made their own assessments and decided to stop buying wool from OVIS21 as they found animal cruelty in 1 of 26 factories. This issue involving Cubreme will be included in future reviews of this case.

References

Allwood J, Laursen S, Malvido de Rodríguez C, Bocken N (2006) Well Dressed? The present and future sustainability of clothing and textiles. University of Cambridge, Institute for Manufacturing, Cambridge

Bader P (2008) On the path to a culture of sustainability: conceptual approaches *Nachhaltigkiet*. Retrieved from http://www.goethe.de/ges/umw/dos/nac/den/en3106180.htm. Accessed 12 Dec 2012

Cataldi C, Dickson M, Grover C (2010) Slow Fashion. Tailoring a Strategic Industry Approach towards Sustainability, master's thesis. Blekinge Institute of Technology, Karlskrona

Dickson MA, Loker S, Eckman M (2009) Social responsibility in the global apparel industry. Fairchild Books, New York

Earley R (2007) The new designers: working towards our eco fashion future. Paper presented at the Dressing Rooms: Current Perspectives on Fashion and Textiles Conference, Oslo, Norway, May 14–16, 2007

Ehrenfeld JR (1999) Cultural structure and the challenge of sustainability. In: Sexton K, Marcus AA, Easter KW, Burkhardt TD (eds) Better environmental decisions—strategies for governments, businesses, and communities. Island Press, Washington, pp 223–244

Farrer J (2011) Remediation. Discussing fashion textiles sustainability. In: Gwilt A, Rissanen T (eds) Shaping sustainable fashion: changing the way we make and use clothes. Earthscan, London, pp 19–33

Fatemi , Fooladi I (undated) Corporate social responsibility and value creation. Working paper (Dalhousie University)

Fletcher K (2008) Sustainable fashion and textiles. Design journeys. Earthscan, London

Fletcher K (2014) Sustainable fashion and textiles: design journey, 2nd edn. Earthscan, London

Fletcher K (2012) Preface. In: Gardetti MA, Torres AL (eds) Sustainability in fashion and textiles: values, design, production and consumption. Greenleaf Publishing, Sheffiel, p ix

Fletcher K, Tham M (2015) Introduction. In: Fletcher K, Tham M (eds) Routledge handbook of sustainability and fashion. Routledge, New York, pp 1–11

Freeman RE (1984) Strategic management: a stakeholder approach. Pitman, Boston

Freeman RE, Martin K, Parmar B (2007) Stakeholder capitalism. J Bus Ethics 74:303–314

Fuad-Luke A (2004–2005) Slow: slow theory. A paradigm for living sustainably? http://www.slowdesign.org/slowtheory.html. Accessed 02 Oct 2011

Gardetti MA (2005) Sustainable development, sustainability and corporate sustainability. In: Gardetti MA (ed) Texts in corporate sustainability: integrating social, environmental and economic considerations with short and long term. LA-BELL, Buenos Aires

Gholami S (2011) Value creation model through corporate social responsibility. Int J Bus Manag 6(9):148–154

Gupta NJ, Benson CC (2011) Sustainability and competitive advantage: an empirical study of value creation. Compet Forum 9(1):121–136

Gwilt A, Rissanen T (2011) Introduction from the Editors. In: Gwilt A, Rissanen T (eds) Shaping sustainable fashion: changing the way we make and use clothes. Earthscan, London, pp 13–14

Hart SL (1995) A natural-resource-based view of the firm. Acad Manag Rev 20(4):986–1014

Hart SL (1997) Beyond greening: strategies for a sustainable world. Harvard Bus Rev 75(1):66–76

Hart SL (2005) Capitalism at the crossroads. Wharton School Publishing, Upper Slade River

Hart SL (2007) Capitalism at the crossroads: capitalism at the crossroads—aligning business, earth, and humanity, 2nd edn. Wharton School Publishing, Upper Salde River

Hart SL, Milstein M (1999) Global sustainability and the creative destruction of industries. MIT Sloan Manag Rev 41(1):23–33

Hart SL, Milstein M (2003) Creating sustainable value. Acad Manag Exec 17(2):56–67

Hawley J (2011) Textile recycling options: exploring what could be. In: Gwilt A, Rissanen T (eds) Shaping sustainable fashion: changing the way we make and use clothes. Earthscan, London, pp 143–155

Hethorn J, Ulasewicz C (2008) Sustainable fashion: why now? A conversation about issues, practices, and possibilities. Fairchild Books and Visuals, New York

Koefoed O, Skov L (unknown year) Sustainability and fashion. In Openwear: sustainability, openness and P2P production in the world of fashion, research report of the EDUfashion project. http://openwear.org/data/files/Openwear%20e-book%20final.pdf. Accessed 9 May 2012

Kozlowsky A, Bardecki M, Searcy C (2012) Environmental impacts in the fashion industry: a life-cycle and stakeholder framework. J Corp Citizensh 45 (special issue on Textiles, Fashion and Sustainability; Gardetti MA and Torres AL eds): 17–36

Lee M (2003) Fashion Victim. Our love-hate relationship with dressing, shopping and the cost of style. Broadway Books, New York

McCann J (2015) Consumer behavior and its importance in sustainability of clothing thing. In: Muthu SS (ed) Handbook of sustainable apparel production. CRC Press, Boca Raton, pp 241–270

Muthu SS (2014) Assessing the environmental impact of textiles and the clothing supply chain. Woodhead Publishing, Cambridge

Oxfam GB, Clean Clothes Campaign and ICFTU (2004) Play fair at the olympics: 45 hours of forced overtime in one week. Oxfam GB, Oxford

Packard V (1963) The waste makers. David Mc Kay Company, New York

Porter M, Kramer MR (2006) Strategy and society: the link between competitive advantage and corporate social responsibility. Harvard Bus Rev 1–15 (reprint)

Porter M, Kramer MR (2011) Creating share value. Harvard Bus Rev. https://hbr.org/2011/01/the-big-idea-creating-shared-value. Accessed 10 Aug 2015

Post JE, Preston LE, Sachs S (2002) Redefining the corporation: stakeholder management and organizational wealth. Stanford University Press, Stanford

Ross RJS (2009) Slaves to fashion: poverty and abuse in the new sweatshop. The University of Michigan Press, Ann Arbor

Sachs S, Rühli E, Kern I (2008) The globalization as challenge for the stakeholder management. Paper presented at Annual Meeting of the Academy of Management (AoM), Anaheim, California

Sachs S, Maurer M (2009) Toward dynamic corporate stakeholder responsibility. J Bus Ethics 85:535–544

Slater K (2000) Environmental impact of textiles: production, processes and protection. Woodhead Publishing Limited, The Textile Institute, Cambridge

Tobler-Rohr MI (2011) Handbook of sustainable textile production. Woodhead Publishing Limited, The Textile Institute, Cambridge

UK Department for Environment, Food and Rural Affairs DEFRA (2008) Sustainable clothing roadmap briefing note December 2007: sustainability impacts of clothing and current interventions. DEFRA, London

Walker S (2006) Sustainable by design: explorations in theory and practice. Earthscan, London

World Commission on Environment and Development WCED (1987) Our common future. Oxford University Press, Oxford

Facets of Indigo: Combining Traditional Dye Methods with State-of-the-Art Digital Print Technology, A Sustainable Design Case

Kelly Cobb and Belinda Orzada

Abstract "If indigo was invented today, we would never approve it." Reflecting on this statement by Andrew Olah related to a 2014 *Just Style* publication on environmental textiles for apparel led our team of six apparel design scholars into a sustainable design challenge. Our study offers a collective model of sustainable design wherein faculty in a university fashion and apparel program combined efforts and talents to develop a solution for reducing the environmental impact of indigo (while retaining aesthetic richness) through the integration of traditional and digital design. A case study method is adopted as a specific, unique, and bounded system (Stake 2008) that frames the creative process so as to capture best practice within a collective design working model. Within this model we analyze historic dye processes and relevant literature, as well as emergent technologies to define criteria for the resulting design output. Qualitative data in the form of observations, as well as our personal reflections as designers and educators, are transcribed and analyzed.

Keywords Collaborative design · Digital printing · Apparel design · Sustainability · Innovation · Textile dyeing · Creative design · Design process

1 Introduction

In 2014, the apparel design faculty of Fashion and Apparel Studies at The University of Delaware received funding to explore digital print technology with students through a product development project. The project "LEUCO STATE" engaged student and faculty designers by pairing tradition with innovation through

K. Cobb (✉) · B. Orzada
Department of Fashion and Apparel Studies, The University of Delaware,
202 Alison Hall West, Newark, DE 19716, USA
e-mail: kcobb@udel.edu

© Springer Science+Business Media Singapore 2016
S.S. Muthu and M.A. Gardetti (eds.), *Green Fashion*,
Environmental Footprints and Eco-design of Products and Processes,
DOI 10.1007/978-981-10-0111-6_2

the theme of indigo dye. The faculty portion of the grant, which is detailed here, focused on skill-building in print technologies, and the opportunity to conduct research along with the students into textile history and traditional dye methods. We explored ways to combine traditional techniques with innovative applications. Through this project we applied design research into emerging print technology through the lens of the tradition of indigo and developed an opportunity for six faculty members to work collaboratively during the textile and apparel design processes.

Significant to the collaboration was the invitation to the apparel design faculty to think, make, and learn together. As colleagues we "collaborate" to solve curricular and departmental issues; yet, in terms of research, our work is often conducted in solitude. Our ways of working are diverse. As design educators, we have much to teach and to learn from one another. This project challenged us to take that opportunity. The faculty design team consisted of Kelly Cobb, Belinda Orzada, M. Jo Kallal, Adriana Gorea, Katya Roelse, and Martha Hall, all professors and instructors of apparel design at the University of Delaware.

1.1 Background/Context

Throughout history, dyes have been valuable trade commodities, but in many cases have led to horrific exploitations of both people and the environment, as occurred in the indigo and logwood trades (Flint 2008). Indigo-dyed products are part of our global culture. Indigo is one of the oldest natural coloring substances used for textiles. The traditional indigo dyeing process and the subsequent garment laundering (dye residues) distress the environment. Emerging digital textile printing technologies are leading to significant paradigm shifts in design processes and print/surface design aesthetics. New modes of application, production, distribution, and consumption are being introduced. The first section of this chapter discusses the components of our project: indigo dyeing and the environment, digital printing, and collaborative design. This is followed by a discussion of the theoretical framework used in our collaborative design process and how our work fits into this framework. Finally, we share reflections of the experience.

2 Importance of Indigo

Indigo is one of the oldest natural coloring substances used for textiles, a prime source of color. Preparations focusing on indigo are found in ancient chemical documents dating back to 300 BC. At certain periods in history, Indigo has wielded more power than firearms. Kings, poets, and protesters have all donned indigo cloth. Tribal chieftains of Africa, the Middle East, and South America wore dark blue indigo-dyed robes as did members of the ruling classes of China, Japan, and Indonesia. On the other hand, traditionally faded blue work clothes have been chosen deliberately as an antiestablishment political statement (Gordon 2011). During the American Revolution, cubes of indigo replaced paper currency (McKinnley 2011). The fact that a majority of the people in a majority of the countries of the world are wearing blue jeans on any given day constitutes global ubiquity (Miller and Woodward 2011). Known as a living color, indigo's own process of decomposition yields its color. The process of imparting color from the indigo plant is so powerful that it has its own creation myth, a Liberian legend about how the properties of indigo were discovered:

> In those days the people grew much cotton and the weavers wove much cloth. All of it was white and they had a hunger for color, especially the blue of the sky, but they did not know how to make that blue go far down and come into the cloth and stay there.... The water spirits come to Asi in a dream. They tell her that for blue to come down to earth and stay, these things are needed: salt, urine, and ashes to live with the leaves of the indigo (Dendel 1995, p. 41).

2.1 Indigo Properties and Process

Indigo dye is produced in a vat process in which chemical reactions, including fermentation, reduction, and oxidation occur. It imparts a distinctive blue to cloth that has inspired people around the world for thousands of years. Unlike most dyes, vat dyes are not water soluble and require a chemical process, activating the dye and thus attracting it to the fiber. The process of converting these dyes to a soluble leuco form requires an alkaline, oxygen-reduced bath. In this bath, the vat dye changes from its insoluble form to its leuco form. In its leuco form, the dye color attaches to the fiber and when the fiber is removed from the dye bath, the dye oxidizes on the fiber, returning to its insoluble form and color (Brackmann 2006, p. 133).

2.2 Natural Versus Synthetic Indigo

Dye is extracted from the indigo plant through a process of fermentation involving plant matter and water with the addition of a caustic agent. Fermentation methods were replaced by synthetic indigo in the last century. Synthetic processes were adopted to accommodate the new scale and speed necessary for mass production. The chemicals used in producing synthetic indigo pigment, and throughout the manufacturing process include aniline, sulphur, sodium hydroxide, hydrosulphate, and formaldehyde, which can be harmful both to humans, through inhalation, and to the environment where they are discharged after dyeing (McGinn 2013).

3 Textile Industry Environmental Impacts

The textile industry is a chemical-intensive industry. After agriculture it is the number one polluter of clean water (Cao et al. 2014). According to a 2005 report by the Hazardous Substance Research Center, an estimated 17–20 % of industrial water pollution comes from textile dyeing and finishing treatments given to fabric. Some 72 toxic chemicals have been identified in water solely from textile dyeing, 30 of which cannot be removed. Considering both the volume and the composition of chemicals in textile effluents, such as the massive presence of dyes, salts, additives, detergents, and surfactants, the textile industry is rated as the most polluting agent among all industrial sectors (Anastas and Zimmerman 2003).

3.1 Pollution and Effluents

Dyeing and finishing one ton of fabric can result in the pollution of up to 200 tons of water according to Glausiusz (2008). Textile mills discharge millions of gallons of effluent each year, full of chemicals such as formaldehyde (HCHO), chlorine, and heavy metals (including lead, mercury, and others) which are significant causes of environmental degradation and human illness. Water samples taken downstream from textile plants in Tehuacan, Mexico, a major denim-producing region, have been shown to contain lead, mercury, cadmium, and selenium. Local farmers complain of chemically burned seedlings and sterile soil (Glausiusz 2008). A 2011 internal wastewater study undertaken by Patagonia (Chouinard and Stanley 2013) found that it takes a textile mill about 500 gallons of water to produce enough fabric to cover a couch. To grow the cotton, then weave and dye the fabric for a single Patagonia pima cotton shirt uses over 600 gallons, the equivalent of a day's drinking water for 630 people. And 15 years from now, between a third and half of the world's population will be living in areas plagued by drought. According to Patagonia, pollution of the Pearl River where it flows into the South China Sea is visible as an indigo color on Google Earth. Indigo is the color of denim and likewise of the discharge from the world's major jeans factories upstream in Xingtang (Ibid.).

3.2 Textile Industry Innovations: Digital Printing

"Much creativity consists of a new combination of existing ideas. Where the existing ideas are present in different people, it requires some kind of interaction to produce the combination" (Langrish 1985, p. 12). Combining computer technology with traditional hands-on design methods has become a common practice in textile and apparel design. Garment patterns are designed through draping or flat pattern methods, then digitized into a CAD system for production. Hand-drawn illustrations are scanned in and enhanced through a computer graphics system.

The idea that computer technology supports or facilitates creativity is supported by many researchers (Bye and Sohn 2010). Polson et al. (2015) note that print-on-demand digital textile printing provides designers, "the potential for economic and creative independence while reinforcing and sustaining cultural identity and craft traditions" (p. 95). In their research on creativity in apparel design, Bye and Sohn explored the use of computer technology and traditional design methods at three different apparel companies. "The personal perceptions of this group of designers indicate a range of feelings regarding how technology and hands on methods influence their creativity. It is a positive promise for the future that the majority believe that they can be creative using both methods, often with a synergistic effect" (Bye and Sohn 2010, p. 215).

"The blurring of the boundaries between production tool and metamedia in the domain of printed textile design is resulting in changes in working processes, development of new hybrid craft techniques and a changing visual language of pattern and colour on cloth" (Tredaway 2004, p. 258). Use of digital printing technology to enhance the textile and apparel design process has been explored by a number of scholars. Campbell and Parsons (2005) share their design method which incorporates photography and personal artwork manipulated using graphic design software, enhancing digital printing with hand embellishment, and manipulation or scaling of the image to fit a garment pattern design.

Digital textile printing has influenced both style and definition of textile printing. Emerging new looks in prints are possible, because design effects can be developed via computer software. Digital printing also provides quicker sampling for mills, and limited runs for small manufacturers and consumers (Ujiie 2001). Digital print-on-demand systems are designed to minimize fabric waste and ink use (Spoonflower 2015). Digital printing allows print-on-demand options for small lengths of fabric, and printing only what is ordered. Therefore, there is no wasted printed fabric. The printing system mixes color as the fabric is printed (from four to six basic colors), as opposed to the screen-printing process which requires colors to be mixed in advance of the print process. Additionally, wet postprocessing is not required in the digital print process therefore no water is consumed during printing.

4 Collaboration

Collaboration is a purposeful relationship (Rubin 2002). This creative and experimental process leads much to more complex design thinking and analysis than is possible with only one designer (Campbell and Parsons 2005). Through collaborative practice there is an opportunity to mesh skills and expertise, offering entry points into meaningful opportunities for shared exploration. Collaborators plan, decide, and work jointly on an activity to generate a shared process that results in the product. According to Poggenpohl and Satō (2009), collaboration is poorly defined because it exists in multiple domains of exploration resulting in diverse outcomes. In an analysis of the varied definitions of collaboration, key characteristics emerge, including the sharing of knowledge, the bridging of disjointed knowledge, and the production of something not otherwise possible, by reacting, cooperating, and participating in a spirit of trust.

Where contribution prevails, participants' roles are more narrowly defined and ownership of process and product autonomous. By contrast, collaboration is interactive in the sense that ownership of specific aspects is relinquished for the sake of an integrated whole. Collaboration involves working together "through shared decision-making, the give and take of ideas exchanged and explored, the integration of multiple perspectives and a synthesis that integrates hitherto isolated or incompatible ideas" (Poggenphol and Satō 2009, p. 142). In terms of design collaboration Block and Nolert (2005) suggest that collaboration should result in an outcome that could not be achieved alone by individuals, and the work in the end reflects a blending of all participants' contributions. A sock, for example, can be viewed as just one tangible result of collaborative efforts of diverse sectors within the apparel supply chain.

4.1 Collaboration in the Apparel Industry

A recent economic survey of 1,656 executives from 100 countries revealed that collaboration is vital to future successes. A defining feature within a model of sustainable business is the movement from traditional top-down hierarchies to collaborative flat models of organization. Abrams (2005) suggests that contemporary work demands collaboration, communication, speed, interaction, teamwork, and creativity and that the new office demands the networking of intelligent autonomous individuals as a prerequisite to problem solving. Regarding the discipline of design, the designer is not the lone genius "outlier," producing solitarily at a specific stage of business procedure, rather she is an equal stakeholder, integrated with management, shaping what organizations have to offer and express (Jevnaker in Poppenpohl and Satō 2009, p. 29) Designers as equal stakeholders that exist beyond their own creative silo are critical to the industry agenda. A case in point is the recent Outdoor Retailer Summer Market (a comprehensive outdoor buyer demographic for the outdoor apparel market). Their 2015 speaker series, "Can't We All Just Get Along? Great Design Requires Collaboration," focused on the theme of collaboration within the industry and generated dialogue with industry stakeholders seeking to demystify the myth of the lone genius.

4.2 Collaboration in the Creative Scholarship of Apparel Design

Business and social science have different viewpoints on collaborative action; business, in large part, attends to product and procedure, as well as to performance and output, whereas social science focuses on individual and group insights that lead to social process (Poggenpohl and Satō 2009, p. 137). Surprisingly, collaboration as a topic of apparel scholarship is not well represented in the literature. Collaboration is described in niche models of design (e.g., Dilys Williams' 2014 reference to BOUDICCA in *Fashion Practice*). In academic projects, collaboration refers to interactions between educators, students, and industry partners (Karpova et al. 2011; Byun et al. 2012).

There is scant research articulating collaborative actions (specific to apparel) among design scholars. Campbell and Parsons (2005) discuss collaborative process within the design process, but they do not define collaboration. As educators of fashion design students we collaborate to solve curricular and departmental issues; we collaborate in the classroom to facilitate learning, yet in terms of research, our work is often conducted alone. The connection here to the myth of the lone wolf is significant but is not elaborated upon except to infer that the notion of the solitary expert creating in an information silo is obsolete in academia as well as business. Collaborative action is the sustainable model moving forward. Academia offers incentives for interdisciplinary interactions with science, the arts, business, and engineering, as the mode of engagement that generates new knowledge while stabilizing the discipline. For the sake of this project, we define collaboration as a shared action involving activities such as sharing, motivation, communication, diversity, support, and problem solving.

5 Theoretical Framework

The theoretical framework for our design process was based on the Design Council's stages of the Double Diamond design process model (http://www.designcouncil.org.uk/news-opinion/design-process-what-double-diamond). The Design Council takes a holistic approach to design, breaking down the design process to four commonalities in the creative process. These are discover, define, develop, and deliver. The visual map of the design process is in the form of a double diamond (Fig. 1). Divergent and convergent thinking are visualized by the broad and narrow areas. Developing a number of ideas or prototypes, then narrowing down the possibilities, happens twice in this iterative process model (Hunter 2015).

The discover stage of the double diamond model is the beginning of the project. Designers explore, research, try out various methods, and keep their eyes open for a broad range of ideas and opportunities. Activities to assist the discover stage are suggested, such as creating a dedicated project space, observation, journaling, brainstorming, being your user, surveys, fast visualization, and more (Hunter 2015). In the define stage, "Designers try to make sense of all the possibilities identified in the Discover phase. Which matters most? Which should we act on first? What is feasible? The goal here is to develop a clear creative brief that frames the fundamental design challenge" (Hunter 2015, np). Activities

Fig. 1 Adaption of double
diamond design model.
(Courtesy of Authors 2014.)

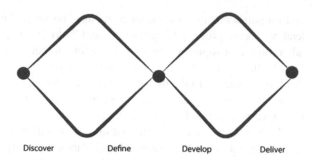

Discover   Define   Develop   Deliver

recommended in this stage include: holding focus groups, and developing assessment criteria. The develop stage provides designers the opportunity to create solutions or concepts, and is the time when prototypes are tested and iterated. It is a trial-and-error process of idea refinement. Developing use scenarios and physical prototyping contribute to the refinement that occurs during the develop stage. Delivery is the final stage of the double diamond model. In this stage the project is finalized and readied for production, or exhibition. Final testing and evaluation of the project occur here. Designers must evaluate whether the project meets the design and use criteria (Hunter 2015).

5.1 The Design Process

The stages of collaborative design experienced by our team followed Salonen's (2012) interpretation of the double diamond model. Salonen frames his discussion of the discover, define, develop, and deliver stages as a framework for designers to explore, test, and innovate in a collaborative or team setting. Ultimately, we developed an open and flat organizational system in which each designer contributed to the process and developed a portion of the final ensemble. A Google blog site was developed to enhance communication between the designers.

5.2 Discover

Initial goals of our project were to explore the indigo dye process, to be inspired through textile dyeing and surface design opportunities, and to create an opportunity for creative scholarship for the apparel design faculty. The additional opportunity for collaboration was very loosely defined in the beginning and evolved over time.

The designers were each at different levels in their experience with dye processes, surface design techniques, and natural dye knowledge. Our exploration of these began by attending an indigo workshop given by master dyer Christina

Roberts at The Fabric Workshop and Museum in Philadelphia, PA. The designers worked alongside master dyers to formulate two vats to test: an organic fructose vat, as well as a chemically reduced vat, from which basic samples were developed.

Following these workshops, the designers individually researched surface design techniques in preparation for a planned dye studio day. Each designer explored inspirational images and researched the history and process of indigo dyeing. Several designers were directed to shibori techniques as surface design methods often used with indigo dyeing. Each designer utilized the blog space to share inspirational images, mind maps, sketches, and thoughts about the early direction of his or her ideas (Fig. 2).

Traditional crafts have often been sources of inspiration and appropriated by Western craftsmen and designers (Hedstrom 2000). Our team of designers likewise was inspired by tradition within indigo dyeing. We applied traditional shibori resist methods to develop multiple surface design samples, testing various combinations of technique, material, and process. Within our group the extent to which traditional techniques were explored varied greatly. Gorea and Kallal were very inspired by shibori techniques. After researching shibori techniques, Gorea found herself attracted to many of them and realized she could "incorporate several wrapping styles into something that can look very simple, but otherwise is

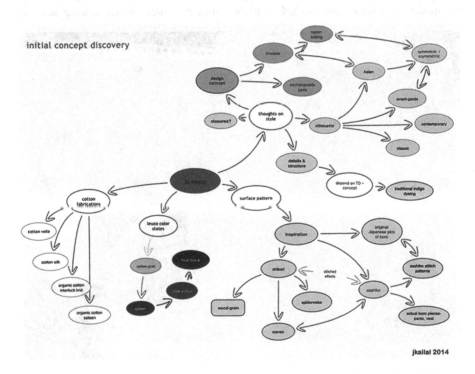

Fig. 2 Concept of discovery inspiration mindmap. (Courtesy of authors 2014.)

very difficult to achieve via traditional techniques." Kallal developed a number of different traditional shibori stitched samples to dye.

Roelse, on the other hand, visualized the traditional techniques as a "process on a pedestal." She sought to keep the "experimental and traditional stage small and contained, and by using these discoveries only as a jumping off point—a seed from which to design new fresh designs—the outcome will only be vaguely reminiscent of its beginning."

5.3 Define

"Collaboration is not a one-dimensional idea. It's the result of using the right tools in a well-thought out process" (Charbin 2010). To facilitate a multidesigner collaboration, we chose to work within the parameters of a clothing "ensemble". Each designer would develop a textile print and a portion of the multicomponent ensemble. The process involved collaborative work sessions in the department textile lab, individual work sessions in personal studios, and interaction within the multiauthored project blog.

As Salonen (2012) notes, collaboration also adds another level of complexity to a project. At different stages in the design process, the structure of the project and the mode of collaboration varied. Collaborators met face-to-face during several

Fig. 3 Preliminary design sketches. (Courtesy of authors 2014.)

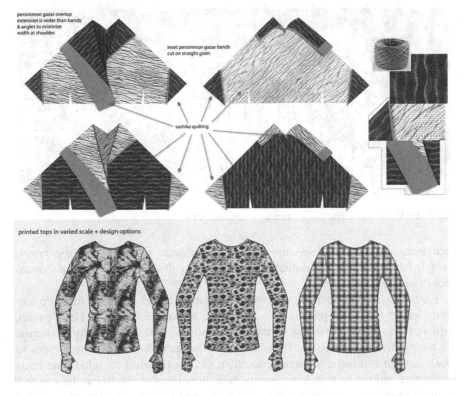

Fig. 4 Garment flat sketches. (Courtesy of Designers 2014.)

site visits and studio days, yet a second, and primary, site of interaction existed virtually on a multiauthored project blog as designers posted regarding inspiration and process and others commented on colleagues' posts. Through these discussions, we shared opinions and worked through design ideas for both our textile design and garment design. Initial sketches of several design ideas were produced (Fig. 3).

Our use of indigo and shibori surface design techniques initially led some designers to Japanese-inspired garment shapes. Further discussion led to each designer sharing his strengths and preferences for developing a certain portion of the ensemble and preliminary decisions on textiles. Kallal and Roelse each used graphics software to scale their shibori prints into garment flat sketches (Fig. 4).

5.3.1 Develop

Developing both textile print and garment designs in a collaborative work and decision process was a key consideration in the design concept and the collaborative process. Indigo-dyed fabric samples using shibori and other surface design

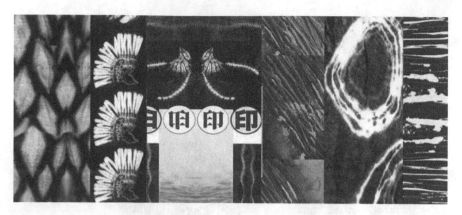

Fig. 5 Digital design variations. (Courtesy of authors 2014.)

techniques were manipulated using graphics software to develop fabric prints. Sample fabric yardage was digitally printed to test color and proportion of the surface design (Fig. 5).

Each designer individually and in collaboration began work on three-dimensional prototype development. Two designers, Kallal (bolero) and Hall (pants), explored zero waste pattern design as an additional sustainable design criterion for their individual garments. They each tested paper and/or muslin prototypes to work through silhouette, fit, and proportion. Orzada (bustier) recycled blue jeans by draping and fitting pant legs directly on a dress form to determine the garment shape. Gorea mathematically planned the pleating process and length of fabric needed for her pleated skirt.

Once we had each reached the prototype stage, we met in the studio to share our work. Visual analysis of the garment prototypes, partially completed garments, and digitally printed sample fabrics allowed us to discuss proportion and final garment details as a group. Cobb had the idea to incorporate another texture in the look by using leather dyed with shibori techniques and Orzada made plans to incorporate a leather strip as the center back lacing of the bustier. Gorea and Cobb collaborated to incorporate Cobb's textile into the skirt design.

There was "give and take" experienced by all during the collaboration process. Hall noted, "[A] danger of working with several talented designers, is that we each want to create something unique and strong. However, speaking for myself, I could see very quickly that I was 'over-designing' and needed to edit my ideas to be in dialog with, rather than fighting with, the other garments in the ensemble." Roelse states, "[T]hinking about my overall approach to the creative process and working in a group, I try and let go of my personal vision and see what is needed instead. It's good practice not to have a 'precious' design."

Gorea found this editing necessary as well; she states "while draping my muslin over Martha's pants I realized I will have to… add fabric fullness." She felt piecing the print in the back would be too predictable, "so I thought of using a strip of the border print vertically to create a wrap like effect (Fig. 6)."

Fig. 6 Image of skirt
prototype. (Courtesy of
authors 2014.)

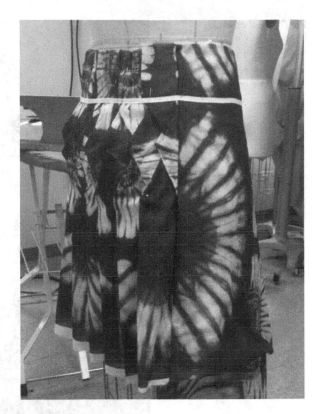

5.4 Deliver

Composed of many prints and garments, the result is a visual, tactile, and dynamic form of conversation. The outcome is a creation of one-off garment pieces (in print and silhouette) that work in conversation with each other as an ensemble. The layering of pattern and garment into one "look" is aesthetically significant. Considering the balance of all pieces, Orzada noted that the rawness of the bustier with its frayed edges, ingrained soil, and hand stitching "provides a great balance to the preciseness of other pieces in the ensemble. The solid background was needed as a relief to all the varied prints, and provided a canvas for using everyone's prints to connect them all."

The design team submitted the ensemble to the International Textile and Apparel Association's 2014 juried Design Exhibition. *Cloud Eater* was awarded the 2014 Educators for Socially Responsible Apparel Practices Award for Sustainable Apparel Design (Figs. 7 and 8).

Fig. 7 Final ensemble
image, front view. (Courtesy
of authors 2014.)

6 Reflections

Designing for sustainability is a daunting task. As designers, the project was an invitation to problem solve issues collectively, experiment with new ways of working, to guide and be inspired by designer collaborators. In working together, the team achieved a sustainable solution that was beyond the scope of individual inquiry.

Through collaborative applied research, the team developed a design solution to reduce the impact of indigo (while retaining the aesthetic richness) through the integration of traditional and digital design. Design scholars developed an applied knowledge of collaborative design that considered the potential of technologies from multiple perspectives. In the collaborative design process, it was important to develop an opportunity for us to be reflective practitioners, engaging in an examination and evaluation of our own work. Establishing an outlet for communication through a blog provided this opportunity. There was varied success to this part of

Fig. 8 Final ensemble image, detail. (Courtesy of authors 2014.)

the design process. Some designers contributed often to the blog, others not as regularly, or in as much detail. This is a potential area for growth in future projects. Deliberate practices could be set up to gain regular insight during the design process.

7 Mapping and Compositing Our Process

In the process of design, collaborator Jo Kallal modified Newman's "design squiggle" to represent her design process. As a team we adopted the squiggle as a format to view our process of codesign. Specifically, our interest was in capturing how each design process aligned and diverged and as a method for visualizing our collaborative design approach (Fig. 9).

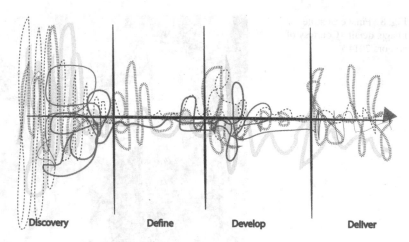

Fig. 9 Facets design squiggle. (Courtesy of Authors 2014.)

7.1 Benefits and Limitations

The Facets collaboration yielded a collection of repeat pattern yardage produced via an "on demand" digital print firm. Innovation of traditional textile pattern creation through resist methods was achieved through digitally iterating smaller scale resist-dyed samples. Our design scholarship involved small batch production of digital textiles, purporting to (1) limit greenhouse gas emissions in comparison with traditional screen printing, as well as (2) generate less production waste (Kujanpää and Nors 2014). Further research (as in a life-cycle analysis) would be necessary to validate claims of sustainability quantitatively. The opportunities and limitations inherent in the scalability of digital printing is another area worthy of further research.

Benefits of the resulting design include retention of the more spontaneous results of hand-dyeing along with the reduction of indigo deposits in water. Innovation of traditional dye methods was achieved with the adoption of digital printing methods, thus lowering the amount of water waste in the dye process. However, digitally printed textiles are pretreated with chemicals, which add to impacts in production. Expanding this initial design study into environmental performance evaluation comparing traditional resist-dye methods with samples digitally integrated into printed yardage would add to the emerging canon of knowledge on digital printing as a sustainable practice.

Digital compositions expand the potential for more quickly developing a complex pattern (i.e., taking a few good "moments" from dye samples and creating patterns in Photoshop rather than producing more yardage using the original technique). Limitations of the design results include the fact that printed fabrics have a white background; this made it difficult to tear/stitch some fabrics or to use the reverse side of the fabric. In our process we missed the tactile qualities of hand dyeing, as some of the complex variations that arise in dye work cannot easily be digitally duplicated. A further study on the aesthetics of traditional hand dye methods in comparison to digital iterations might also prove consequential to the growing field.

Collaboration as a topic of apparel scholarship is not well represented in the literature. We have much to learn by engaging in design with others from different backgrounds within design, different interests and working methods at different stages in one's career. In terms of the collaborative working model, our designers reflected that it was an intriguing opportunity to use like-minded, yet diverse backgrounds, experiences, and perspectives to produce an eclectic look. Each designer was able to gain new skills and experiences, while at the same time draw on her individual backgrounds, textile design, and pattern-making skills to produce results that reflected strong outcomes both individually and collectively. Their contributions balanced their foundation of strengths as apparel designers, and the collaboration required that we push those skills forward and acknowledge the skills of others.

References

Abrams R (2005) All together now: communication networks and collaborative spaces. http://www.adobe.com/motiondesign/MDC_Think_Tank.html. Accessed 15 July 2015

Anastas PT, Zimmerman JB (2003) Design through the 12 principles of green engineering. Environ Sci Technol 37:94A–101A

Block R, Nollert M (eds) (2005) Collective Creativity (exh. cat.) Kassel, München, Frankfurt am Main, Kunsthalle Fridericianum Siemenas Arts Program, Revolver

Brackmann H (2006) The surface designer's handbook: dyeing, printing, painting, and creating resists on fabric. Interweave Press, Loveland Co

Dye E, Sohn M (2010) Technology, tradition, and creativity in apparel designers: a study of designers in three US companies. Fashion Practice 2(2):199–222

Byun S, Kim H, Duffey M (2012) A Multicourse collaborative project within a global context: multidimensional learning outcomes for merchandising and interior design majors. Clothing Text Res J 30(3):200–216

Campbell JR, Parsons J (2005) Taking advantage of the design potential of digital technology for apparel. J Text Apparel Technol Manage 4(3):1–10

Cao H, Scudder C, Howard C, Piro K, Tattersall H, Frett J (2014) Locally produced textiles: product development and evaluation of consumers acceptance. Int J Fashion Des Technol Educ 7(3):189–197

Charbin A (2010) Collaboration in the fashion industry. Case Study. http://www.fibre2fashion.com/industry-article/46/4510/collaboration-in-the-fashion-industry1.asp. Accessed 30 March 2014

Chouinard Y, Stanley V (2013) The responsible company: what we've learned from patagonia's first 40 years. Patagonia Books, New York

Dendel EW (1995) You cannot unsneeze a sneeze, and other tales from Liberia. University Press of Colorado, Niwot

Flint I (2008) Eco colour: botanical dyes for beautiful textiles. Murdoch Books, Millers Point

Glausiusz J (2008) How green are your jeans? http://archive.onearth.org/article/how-green-are-your-jeans. Accessed 18 July 2015

Gordon B (2011) Textiles: the whole story: uses, meanings, significance. Thames and Hudson, London

Hazardous Substance Research Centers/South and Southwest Outreach Program (2005) Environmental hazards of the textile industry. Environmental Update #24, Business Week

Hedstrom AL (2000) Shibori: tradition and innovation. In: Textile society of America symposium proceedings. Paper 816. http://digitalcommons.unl.edu/tsaconf/816

Hunter M (2015) The design process: what is the double diamond? http://www.designcouncil.org.uk/news-opinion/design-process-what-double-diamond. Accessed 13 July 2015

Just Style (2014) Sustainable textiles for apparel: fact, fiction and future prospects report description, p. 1. http://www.just-style.com/market-research/sustainable-textiles-for-apparel-fact-fiction-and-future-prospects_id192005.asp

Kapova E, Jacobs B, Lee JY, Arnold A (2011) Preparing students for careers in the global apparel industry: experiential learning in a virtual multinational team-based collaborative project. Clothing Text Res J 29(4):298–313

Kujanpää M, Nors M (2014) Environmental performance of future digital textile printing (VTT-CR-04462-14). Espoo: VTT Technical Research Centre Finland. http://www.vtt.fi/inf/julkaisut/muut/2014/VTT-CR-04462-14.pdf

Langrish J (1985) Innovation management: the role of creativity. Institute of Advanced Studies Manchester Polytechnic, Manchester

McGinn E (2013) Natural vs synthetic indigo dyes. http://source.ethicalfashionforum.com/article/natural-vs-synthetic-indigo-dyes. Accessed 18 July 2015

McKinley CE (2011) Indigo: in search of the color that seduced the world. Bloomsbury, New York

Miller D, Woodward S (2011) Global denim. Berg, Oxford

Poggenpohl SH, Satō K (2009) Design integrations: research and collaboration. Intellect, The University of Chicago Press, Chicago

Polston K, Parrillo-Chapman L, Moore M (2015) Print-on-demand inkjet digital textile printing technology: an initial understanding of user types and skill levels. Int J Fashion Des Technol Educ 8(2):87–96. doi:10.1080/17543266.2014.992050

Rubin H (2002) Collaborative leadership: developing effective partnerships in communities and schools. Corwin Press, Thousand Oaks, Calif

Salonen E (2012) A designer's guide to collaboration. http://www.designingcollaboration.com/. Accessed 31 March 2014

Spoonflower (2015) Are Spoonflower products eco-friendly? https://support.spoonflower.com/hc/en-us/articles/204465960-Are-Spoonflower-products-eco-friendly. Accessed 17 July 2015

Stake RE (2008) Qualitative case studies. In: Denzin NK, Lincoln YS (eds) Strategies of qualitative inquiry, 3rd edn. Sage, Thousand Oaks, p 112

Treadaway C (2004) Digital Imagination: the impact of digital imaging on printed textiles. Textile J Cloth Cult 2(3):256–273

Ujiie H (2001) The effect of digital textile printing technology on textile design styles. Downloaded June 15, 2015 from http://www.techexchange.com/index_libraryTE_articles

Williams D (2014) Fashioning the future. Fashion Pract J Des Creative Process Fashion Ind 6(1):131–140

Understanding Consumer Behavior in the Sustainable Clothing Market: Model Development and Verification

Małgorzata Koszewska

Abstract The main purpose of this study is to expand the knowledge of consumer behavior in the market for sustainable clothing and to build a theoretical model of this behavior based on the review of the relevant literature and the author's own research. The model is to provide a wider perspective on consumer behavior with respect to textile and clothing products with ecological and social characteristics, as well as describing the relationship between the consumer and the manufacturer of textiles and clothing pursuing a strategy founded on the principles of corporate social responsibility (CSR). The selected elements of the theoretical model are verified empirically by means of structural equation modeling (SEM) using a representative sample of 981 Polish customers. The research findings show that consumers' attitudes towards apparel shopping have a significant and positive influence on their willingness to pay a premium for sustainable products, on the recognizability of ecological and social labels, and, finally, on the actual purchase of sustainable clothing. These results give additional evidence pointing to a prominent role of the recognizability of ecological and social labels in purchasing sustainable apparel. The results of the study allow better understanding of factors determining consumer behavior towards sustainable clothing and suggest practical solutions to their producers.

Keywords Consumer behavior · Model · Sustainable consumption · Structural equation modeling · Clothing · Green fashion

M. Koszewska (✉)
Department of Materials and Commodity Sciences and Textile Metrology,
Team of Market Analyses of Product Innovations, Lodz University of Technology,
Ul. Zeromskiego 116, 90-924 Lodz, Poland
e-mail: malgorzata.koszewska@p.lodz.pl

© Springer Science+Business Media Singapore 2016
S.S. Muthu and M.A. Gardetti (eds.), *Green Fashion*,
Environmental Footprints and Eco-design of Products and Processes,
DOI 10.1007/978-981-10-0111-6_3

1 Introduction

Despite significant theoretical and empirical advancements of studies into consumer behavior and growing interest in sustainable development and corporate social responsibility (CSR; Ogrean and Herciu 2014), studies exploring the textile and clothing market are still few. In the last few years, though, the number of studies focused on consumers' attitudes towards sustainable textile and clothing products has clearly increased interest and new models of consumer behavior in the market for these products have been developed. Yet, most of them address only some selected aspects of consumer behavior rather than attempting to present a comprehensive picture of the relationships between producers and buyers. This chapter has been designed with a view to reducing this gap.

There are several reasons why consumer behavior in the sustainable clothing market should deserve more attention from researchers.

First, gradual environmental degradation, shrinking of nonrenewable resources, lower quality of life, and social and ethical problems are directly or indirectly arising from snowballing consumption and in the case of textiles and clothing sector from the fast-fashion trend. These unfavorable processes will not stop unless consumption patterns are modified. Manufacturers may use new designs and technologies to minimize the impact of a product on the environment and to make production more sustainable but their efforts are pointless if consumers do not buy more sustainable goods and do not change their consumption habits. Therefore analysis of immediate connections and influences of the demand side of the market (consumers) and the supply side of the market (producers) is crutial (Koszewska 2011b; Muthu et al. 2013).

Textile and clothing products accompany people "from cradle to grave," giving them comfort and protecting their health. Although many industries are faced with the challenges of sustainable development, few of them are exposed to demands from consumers, the media, and NGOs to act on the principles of sustainability as much as the textile and clothing industry is. The author is of the opinion that for the demands to be met consumer behavior and its determinants must be thoroughly analyzed.

Accordingly, this study has been designed to review factors determining consumer behavior towards sustainable textiles and clothing[1] and to develop and test a model of this behavior.[2]

---

[1]In the chapter, sustainable textiles and clothing products are meant to be products that are:
- Manufactured in a socially responsible manner, that is, in a way satisfying the economic requirements and addressing the environmental and social aspects. These innovative products naturally contribute to sustainable development, and are
- Made to be used by individuals; this means that clothing represents the bulk of them but home textiles such as carpets, curtains, bedclothes, tablecloths, towels, and the like are also significant. For convenience, a collective term of sustainable clothing is used in this chapter.

[2]For the purposes of this study, customer behavior is understood as all actions and perceptions of an individual causing his or her to want to buy a product, to choose a product, and finally to buy it.

When a new model of consumer behavior is being created or one that already exists is being adapted, the special characteristics of the analyzed market or of its segment must be taken into consideration. This is important, because the socially responsible behavior of consumers may differ significantly depending on the type of goods (McDonald et al. 2009) and because fashion consumers differ from other consumers in ethical consumption decision making (Niinimäki 2010). Last, models dedicated to particular markets are more likely to be used by firms for business purposes.

The structure of this chapter is the following. Section 2 discusses the classification of consumer behavior models and positions the model proposed by the author among the existing ones. Section 3 presents a comprehensive theoretical model of consumer behavior towards sustainable clothing based on a literature review. The model describes relationships between consumers and the manufacturers of sustainable products. The model has three theoretical dimensions, namely:

- The supply side: sustainable production systems and sustainable clothing they deliver
- Barriers to the purchase of sustainable clothing
- The demand side: sustainable consumption models

They are discussed in Sects. 4–12, respectively.

Section 13 focuses on the empirical verification of the demand side of the model and on the factors that significantly contribute to the purchase of sustainable clothing. The chapter ends with a discussion, final conclusions, and the limitations of the research.

2 Consumer Behavior Models: Theoretical Background

Consumer behavior models are crucial to disentangling the complexity of relations between consumers and the market, and to understanding how consumers make decisions. They are designed as the simplified representations of consumers' real behaviors, so is it quite natural that they contain many assumptions and suffer from many limitations (Kieżel 2010). The literature on this subject provides many models explaining consumer behavior in the market, and consequently many classifications of the models created based on different criteria (Smyczek and Sowa 2005):

- Simple and complex models
- Structural, stochastic, and simulation models
- Descriptive, prognostic, and normative models
- Theoretical and empirical models
- Static and dynamic models
- One- and multifactor models
- Verbal, schematic, and mathematical models

The above list is not complete. The set of the classification criteria is substantial, therefore the list of the types of consumer behavior models can be much longer than the one above. At the same time, the lines between particular classifications are blurred.

Bettman and Jones have divided consumer behavior models into four broad categories:

- Information processing models of consumer choice
- Stochastics models
- Experimental and other linear models
- Large system models (Bettman and Jones 1972)

Considering the purpose of this chapter, particularly noteworthy are information-processing models that are very frequently employed to analyze consumer behavior towards sustainable products, and large system models. Bettman and Jones have described the large system models as models with a broad general structure of postulated interrelationships, with a somewhat simplified formal model fit within the framework (Bettman and Jones 1972).

The proposed model, too, presents a wide panorama of the relations between producers and consumers. Considering their complexity, only selected elements of decision-making processes performed by customers will undergo empirical verification. The empirical verification is based on the wealth of theoretical and empirical knowledge of structural modeling with latent variables.[3]

A landmark model in the development of structural modeling with latent variables is the Howard–Sheth model. Its authors proposed a model containing latent variables (i.e., theoretical, unobservable variables) and the rules for relating them to their observable indicators. The Howard–Sheth model initiated a series of studies into the formation of relations between the amount of information available and a buying decision that is crucial in analyzing consumer behavior towards sustainable products.

The introduction of latent variables into consumer behavior models allowed accounting for personal values, beliefs, attitudes, and behavioral intentions (the values–attitudes–behavior models), as well as for controlling various mediation and moderation aspects (Sagan 2011).

Jeff Bray has named five major theoretical approaches used in consumer behavior studies (Bray 2008):

- Economic man
- Psychodynamic
- Behaviorist

---

[3]Models with latent variables are one of the main types of models that have been used for more than 30 years now in consumer behavior studies. They owe their popularity to the fact that they allow for considering latent variables representing the unobservable characteristics of consumers and for analyzing the regressive dependencies between them.

- Cognitive
- Humanistic

Each of them posits alternate models of consumer behavior and emphasizes the need to examine quite different variables (Foxall 2004).

The majority of empirically validated models described in the literature on consumer behavior and sustainable products refer to the cognitive orientation, which is closely associated with the already mentioned information processing theory, a basic system of reference for structural modeling in the study of consumer behavior developed in the 1980s. The cognitive orientation encompasses, inter alia, the theory of reasoned action (TRA; Ajzen and Fishbein 1980) and the theory of planned behavior (TPB; Ajzen 1991). These classic theories have been extensively adopted to create models explaining consumer behavior towards sustainable clothing. Many of the models have been verified empirically in the framework of structural equation modeling (SEM), for example, the model of environmentally sustainable textile and apparel consumption (Kang et al. 2013), the model of intention to purchase personalized fair trade apparel (Halepete et al. 2009b), the model of consumer choice in sweatshop avoidance (Shaw et al. 2007), or the model of green purchasing behaviors of urban residents in China (Liu et al. 2012).

The use of the cognitive orientation and the SEM technique for the empirical testing of the models of consumer behavior towards sustainable clothing are discussed more specifically in the next section of this chapter.

The literature contains models that have been created to describe some aspects of consumer behavior towards textile and clothing products (Eckman et al. 1990; Shim and Kotsiopulos 1992; Lea Wickett et al. 1999; Visser and Du Preez 2001; Du Preez 2003; Du Preez and Visser 2003; de Klerk and Lubbe 2008). There are also empirical studies investigating sustainable consumption behavior with respect to textiles and clothing (Butler and Francis 1997; Kim and Damhorst 1998; Blowfield 1999; Shaw et al. 2007; Brosdahl and Carpenter 2010; Ha-Brookshire and Norum 2011; Kang et al. 2013; Hassan et al. 2013; Halepete et al. 2009a; Chan and Wong 2012; Hyllegard et al. 2012; Lee et al. 2012; Han and Chung 2014). However, none of them addresses the ecological, social, or ethical aspects of consumer behavior, nor does it comprehensively explain the extremely complex relationships between producers and consumers. See Fig. 1.

The proposed model is an attempt to provide a more efficient explanation of consumer behavior towards sustainable clothing and to give insight into the relationship between its producers and consumers.

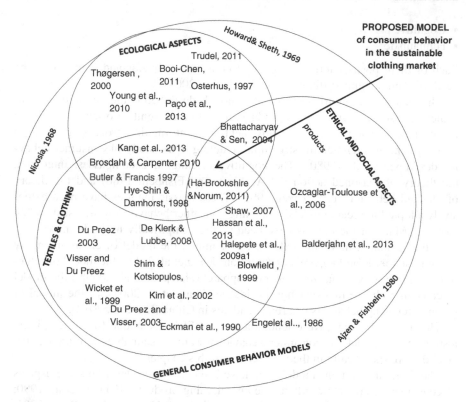

Fig. 1 The position of the proposed model among the existing models of consumer behavior. *AU* Bold label should read "Textiles and Clothing" for consistency (no &). *Source* Developed by the author

3 Theoretical Model of Consumer Behavior in the Sustainable Clothing Market: Relationship Between Consumer and Manufacturer

In this section, a theoretical model of consumer behavior towards sustainable clothing derived from a literature review is presented. This holistic model offers a comprehensive explanation of the relationship between the consumer and the manufacturer of this type of clothing. The basic assumptions underpinning the model have been formulated following an analysis of relationships included in other theoretical models and using the conclusions from the verification of empirical models.

Table 1 presents an overview of the existing consumer behavior models, starting with general models through models accounting for ecological or social aspects of consumption and models of consumer behavior towards textile and clothing products to models explaining consumer behavior towards sustainable clothing.

Table 1 Overview of the evolution of consumer behavior models

Author	Constructs	Area\Product type	Subject	Empirical verification
General consumer behavior models				
Nicosia (1968)	Company's attributes (message exposure)	General	Structure of consumer behavior; a summary flow chart	Partial
	Consumer's attributes, attitude			
	Search evaluation			
	Motivation			
	Decision (action), purchasing behavior			
	Consumption storage, experience			
	Feedback			
Howard and Sheth (1969)	Inputs—stimulus display (significative, symbolic, social	General	Structure of consumer behavior; a summary flow chart	Partial
	Perceptual constructs (over search, stimulus ambiguity, perceptual bias)			
	Learning constructs (confidence, attitude, choice criteria, motives, intention, brand comprehension, satisfaction)			
	Outputs (attention, brand comprehension, attitude, intention, purchase)			
Engel et al. (1986)	Input (stimuli from marketers)	General	Structure of consumer behavior; a summary flow chart	Partial
	Information processing (exposure, attention, comprehension, acceptance, retention)			
	Memory			
	Decision-making process (problem recognition, search, alternative evaluation, choice, purchase)			

(continued)

Table 1 (continued)

Author	Constructs	Area\Product type	Subject	Empirical verification
	Decisional variables (beliefs, motives, attitudes, lifestyle, intentions, evaluative criteria, normative compliance)			
	Environmental influences (cultural norms, social class, reference group, family, unexpected circumstances)			
Ajzen and Fishbein (1980)	Beliefs about the consequences of behavior and evaluations of the consequences	General	Theory of reasoned action—TRA	Yes
	Attitude towards behavior			
	Beliefs about perceptions of others and motivation to comply			
	Subjective norms about the behavior			
	Behavioral intention			
	Behavior			
Ajzen (1991)	Behavioral, normative, and control beliefs	General	Theory of planned behavior—TPB	Yes
	Attitude towards behavior			
	Subjective norm			
	Perceived behavioral control			
	Behavioral intention			
	Behavior			

(continued)

Table 1 (continued)

Consumer behavior models accounting for ecological and/or social aspect of consumption

Author	Constructs	Area/Product type	Subject	Empirical verification
Osterhus (1997)	Consequences	Energy conservation behavior	Integrated model blending normative, economic, and structural influences	Yes
	Responsibility			
	Social norm			
	Cost reward			
	Personal norm			
Thogersen (2000)	Consumer motivation (motivational antecedents: effectiveness, attitude, belief, trust)	Eco-labeled products	Causal path model explaining consumers'	Partial
	Consumer knowledge		Propensity to pay attention to ecolabels	
	Product availability			
	Paying attention to ecolabels			
	Decision to buy			
Bhattacharya and Sen (2004)	Input: CSR activity (type, investment)	General—CSR activity	CSR framework—consumer-centric conceptualization of CSR	No
	Internal outcomes			
	Company (awareness, attributions, attitude, attachment)			
	Consumer (well-being),			
	Issue/cause(awareness, attitude)			
	External outcomes			
	Company (purchase, price premium, loyalty, resilience)			
	Consumer (behavior, modification)			
	Issue/cause (support)			

(continued)

Table 1 (continued)

Author	Constructs	Area\Product type	Subject	Empirical verification
Ozcaglar-Toulouse et al. (2006)	Attitude Subjective norm Perceived behavioral control Ethical obligation Self-identity Intention Behavior	Fair trade grocery products	Modified model of planned behavior	Yes
Young et al. (2010)	General green values Green criteria for purchase Barriers/facilities Product purchase Feedback	Green technology products	Green consumer purchasing mode	No
Trudel (2011)	Company's CSR actions Consumer attitudes Consumer intentions Consumer actions (buy, pay more, punish)	Products with social or environmental attributes	Model of socially conscious consumerism	No
Booi-Chen (2011)	Personal values Perceived consumer effectiveness (PCE) Environmental attitude Green buying attitude Green buying behavior	Green products	Model to study green buying behavior extended from V-A-B (value attitude behavior)	No

(continued)

Table 1 (continued)

Author	Constructs	Area\Product type	Subject	Empirical verification
Paço et al. (2013)	Man-nature orientation Generativity Attitudes Conservation behavior Environmental friendly buying behavior	Environmentally friendly products	Green consumer behavior model	Yes
Balderjahn et al. (2013)	Consciousness for fair consumption (CFC) Ecological concern (EC) Moral reasoning (MR) Buying fair-trade products	Fair-trade products	Consciousness for fair consumption	Yes
Lavorata (2014)	Retailers' commitment to sustainable development as perceived by consumers Ethical consumption Perceived control Subjective norms Behavioral intention (boycott intention) Behavior (boycott) Retailers' image Loyalty	Hypermarkets' commitment to sustainable development	Influence of retailers' commitment to sustainable development on consumer loyalty and boycotts, based on theory of planned behavior	Yes
Consumer behavior models for textiles and clothing products				
Du Preez (2003)	Market-dominated variables (4P) Consumer-dominated variables (demographics, socio-cultural influences, psychological field) Market and consumer interaction (shopping orientation, previous experience, patronage behavior)	Apparel	A conceptual theoretical model: a macro-perspective, variables influencing apparel shopping behavior	No

(continued)

Table 1 (continued)

Author	Constructs	Area\Product type	Subject	Empirical verification
Du Preez and Visser (2003)	Market-dominated variables (product, place)	Apparel	A theoretical model of selected variables influencing female apparel shopping behavior in a multicultural consumer society	No
	Consumer-dominated variables (demographics, sociocultural influences, life style, culture)			
	Market and consumer interaction (shopping orientation, patronage behavior)			
	Decision-making process			
	Female apparel shopping behavior			
	Tree clusters of female apparel shoppers			
De Klerk and Lubbe (2008)	Apparel consumer (senses, emotions, cognitive)	Apparel	Conceptual framework for the role of aesthetics in consumer apparel behavior	No
	Apparel product (formal qualities)			
	Aesthetics (sensory, emotional, cognitive)			
	Perception of quality			

Consumer behavior models for sustainable clothing*

Author	Constructs	Area\Product type	Subject	Empirical verification
Shaw et al. (2007)	Intention	Sweatshop apparel	Empirically validated model of consumer choice in sweatshop avoidance	Yes
	Attitude			
	Subjective norm			
	Perceived behavioral control			
	Desire			
	Intention			
	Plan			
Brosdahl and Carpenter (2010)	Knowledge of environmental impacts	Environmentally friendly textile and apparel	Model of environmentally friendly consumption behavior	Yes
	Concern for the environment			
	Environmentally friendly consumption behavior			

(continued)

Table 1 (continued)

Author	Constructs	Area\Product type	Subject	Empirical verification
Ha-Brookshire and Norum (2011)	Willingness to pay for socially responsible products	Organic cotton shirts	Research conceptual model	Yes
	Product evaluative criteria	sustainable cotton		
	Demographic characteristics	US-grown cotton shirts		
	Apparel product evaluative criteria			
Chan and Wong (2012)	Product-related attributes (PRA)	Fashion clothing	Model of consumer eco-fashion consumption	Yes
	Store-related attributes (SRA)			
	Price premium level of eco-fashion (PP)			
	Eco-fashion consumption decision (ECD)			
Hassan et al. (2013)	Antecedents (complexity ambiguity, conflict credibility source)	Ethical clothing	Conceptual model of uncertainties in decision-making processes	No
	Uncertainty (knowledge, choice, evaluation)			
	Outcomes (delayed purchase, compromised beliefs, negative emotions)			
Kang et al. (2013)	Consumer knowledge	Environmentally sustainable textile and apparel	Structural model of environmentally sustainable textile and apparel consumption	Yes
	Perceived consumer effectiveness			
	Perceived personal relevance			
	Attitude			
	Subjective norm			
	Behavioral control			
	Behavioral intention			

*More models for sustainable clothing are described in more detail in Sect. 3.3

Source Developed by the author

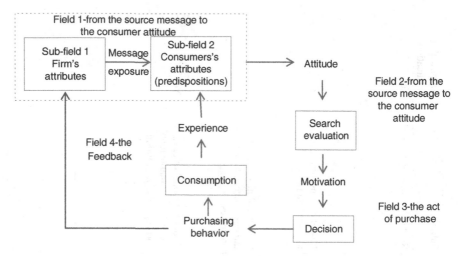

Fig. 2 Nicosia model. *Source* Nicosia (1968)

The starting point for the construction of the proposed theoretical model of consumer behavior towards sustainable clothing was the Nicosia model (1966), one of the first and most popular models created in this field.

There are several reasons for adopting Nicosia's model. First, it focuses on the relationship between the producer and potential customers (Fig. 2). Second, this complex structural model is relatively simple and transparent and it has already been used to analyze relations between consumers and the clothing company (Vignali 1999).

The proposed theoretical model is similar to that developed by Nicosia in that it, too, focuses on the producer, the customer, and the linkages between them. It is different, though, in that it addresses elements that Nicosia has omitted and which, according to the reviewed literature, are relevant to the analyzed type of product (clothing) and its specific characteristics (ecological and social).

The proposed model centers on sustainable clothing, an innovative product containing significant added value represented by its ecological and social characteristics. The consumer evaluation of sustainable clothing versus conventional products leading to its purchase or rejection depends on:

- Producer attributes: Reputation, business behavior and production systems
- Product attributes: Its emotional and use/functional value, price, availability, the place and circumstances of the purchase
- Barriers discouraging a customer from making a purchase
- Consumer attributes and factors determining his or her behavior

All these sets of factors are important for the proposed model (Fig. 3). They are discussed more in detail in the next sections (producer and product, 3.1; barriers, 3.2; consumer, 3.3).

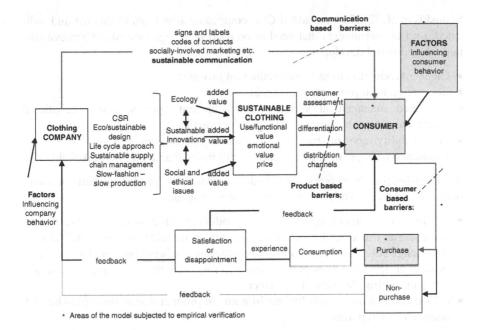

Fig. 3 Theoretical model of consumer behavior towards sustainable clothing: the relationship between the consumer and the producer. *Source* Developed by the author

The producer can influence consumer purchase decisions by means of an integrated CSR communication system, but consumers' decisions depend also on other factors (see Sect. 3.3), which may make them buy or abandon the purchase of sustainable clothing. In both cases, the producer receives feedback and a consumer deciding to buy and use a product gains experience that determines his future decisions (Fig. 3).

4 Production of Sustainable Clothing: Supply Side of the Model

The literature offers a wealth of studies dealing with sustainable production models. Among the methodologies used for developing sustainable products there is integrated eco-design decision making (IEDM) presented by Romli et al. The methodology has three stages (Romli et al. 2015):

- Life-cycle assessment
- An eco-design process composed of three modules (manufacturing, product usage, and end-of-life strategy)
- An enhanced eco-design quality function deployment process

Staniskis et al. (2012) have added CSR communication tools to this list and indicated a set of instruments that need to be used on a regular basis to achieve sustainable industrial development:

- Cleaner production to improve production processes
- Eco-design to improve product characteristics
- Integrated management systems (environmental, quality, and occupational health and safety) to improve management practices
- Sustainability reporting based on sustainability performance
- Evaluation to improve communication with internal and external stakeholders

Additionally, they have proposed a model of sustainable production and consumption as a system. The critical elements of the system are:

- A life-cycle approach optimizing the production process to minimize energy and material use as well as waste output, and to eliminate the "rebound" effect
- Eco-design referring to a systematic incorporation of environmental aspects into product design and development aiming to minimize the environmental impacts along the entire life cycle of a product
- CSR communication including environmental product declarations (eco-labels), sustainability reporting

In the case of textile and clothing the eco-design must meet basic sustainable design principles, such as sustainable fibers, low-impact materials, renewable sources of energy, energy and water efficiency, biomimicry, biodegradable products, pollution prevention, 4R (reduce, recycle, reuse, rebuy), service substitution, community culture, DIY (do-it-yourself) and patchwork, fair trade/ethical practices, and near-sourcing (Thilak and Saravanan 2015).

All these solutions were taken account of in developing the assumptions for the proposed theoretical model of consumer behavior towards sustainable clothing (Fig. 3).

The above list of solutions towards sustainable production systems lacks one instrument that is crucial to the textile and clothing industry: sustainable management of supply chains. The empirical model proposed by Gimenez Sierra points out that supplier assessment and supplier collaboration improve the company's environmental performance (Gimenez and Sierra 2013). Moreover, companies with active sustainability strategies for supply chains demonstrate higher levels of implementation of supplier assessment and of collaboration with suppliers, and therefore higher environmental performance.

Production of sustainable clothing is a very complex process, because the clothing industry has one of the longest, most complicated, dispersed, and geographically stretched supply chains. Compared with the food industry, its manufacturing processes are also less transparent. As a result, understanding and defining the sustainability of textiles and clothing is an ambiguous and relatively problematic exercise. Sustainability is usually associated with the environment, albeit two other aspects of it—economic and social—are as important. Considered in terms of the social impacts of a product, sustainable clothing is a product made by adult workers in good working conditions, and not by children, in line with fair trade rules. However, an objective measurement of social impacts that make sustainable products different from other goods in the clothing market poses some problems.

The sustainability of textiles and clothing is also inextricably linked to fashion. The simple understanding of those two terms indicates a conflict between them as fashion is about being trendy, up-to-date, and the latest whereas sustainability is about long-lasting and durable, low impact, and eco-friendly (Thilak and Saravanan 2015). This conflict is reflected in two opposing fashion trends:

- *Fast fashion* is connected with rapid production, short lead times, increasing number of fashion seasons, lower cost materials and labor promoting overconsumption of low price and low quality garments worn only a few times and discarded quickly.
- *Slow fashion* is connected with slow production that does not exploit natural and human resources and promotes slow consumption of better quality and durability garments bought less frequently but used longer.

In this context, a major challenge for sustainable production systems is to prolong the clothing's life cycle and maximize its utility and functional properties, to encourage slow consumption. Sojin Jung and Byoungho Jin also stress that in slow and sustainable fashion systems quality includes not only the physical, but also design aspects (Jung and Jin 2014). Therefore, sustainable apparel should be characterized not only by durable, easily disposed of materials but also by designs less influenced by fashion trends and multifunctional, which can be worn for a long time, regardless of fashion seasons.

A vital aspect of sustainable production systems is also the *eco-functional assessment* proposed by Muthu, Subramanian, Senthilkannan, et al. (Muthu et al. 2013). This aspect is especially important in the case of textile and clothing products, because according to the research most consumers still tend to apply simple visual criteria to them, clearly focused on the product's benefits for the consumer and its functional properties. Aspects of products such as environmental performance and the rights of workers who make them are verbally stated as important, but daily buying decisions are rarely influenced by them. Only the better off and more aware consumers take them into account. This observation does not apply to aspects that bring immediate functional benefits (Koszewska and Treichel 2015).

The literature offers a whole range of indicators showing the degree of compliance between a product and sustainability criteria (Krajnc and Glavič 2003). They are practically useless for an average consumer, though, because a substantial amount of specialist knowledge is necessary to understand, interpret, and use them. For consumers, much more convenient and useful indicators of the ecological and/or social impacts of clothing are ecological signs and labels (Koszewska 2010, 2011a, 2015). This means of communication between the producer and the consumer has therefore been given a prominent place in the proposed theoretical model. The significance of CSR communication and of its influence on consumers' intentions and buying decisions is covered in more detail and verified empirically in the last section of this chapter.

Fig. 4 Aspects of sustainable development and areas of innovation in the textile and clothing industry. *Source* Developed by the author

A company with an integrated CSR communication system can influence customers, their knowledge, and ecological awareness, as well as their perception of its products. At the same time, the way the company behaves is determined by various external factors, such as NGOs, the policies of the state, consumer choices, and the like (Bhattacharya and Sen 2004). It is worth noting here that although all industries are being forced into compliance with the principles of sustainable development, only few of them experience pressure from consumers, the media, and NGOs as strong as the textile and clothing industry does. It is also important to remember that the recent trends in consumption and consumer behavior clearly respond to the challenges of sustainable development and show along which lines the market for textile and clothing products should develop (Figs. 4 and 5) (Koszewska 2012c).

The future of the textile and clothing industry apparently lies in sustainable innovations such as mass customization (as a counterpoise to fast fashion); energy and raw-material saving technologies; eco-textiles; smart textiles for the elderly, sick persons, and persons with special needs; biodegradable textiles; and so on (Fig. 5).

A sustainable clothing product, its characteristics, price, and the channels of distribution and communication that are the main ingredients of the Kotler marketing mix described in classical decision-making theories link the left-hand side and the right-hand side of the proposed model (supply and demand; Fig. 2).

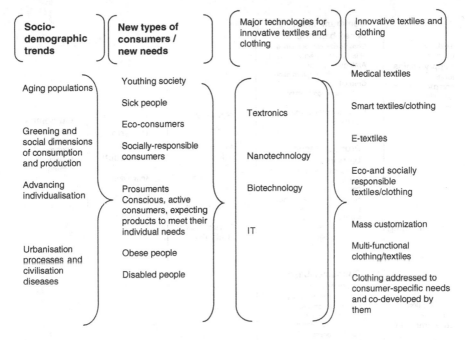

Fig. 5 Sociodemographic trends and directions for the development of textile and clothing innovations. *Source* Koszewska (2012a)

5 Barriers, Perceived Risk, and Uncertainties

The relations between the producer and the consumer may be obstructed by different barriers, causing the latter to delay or withdraw from the purchase of a sustainable product, to buy an unsustainable one, or to feel negative emotions (Hassan et al. 2013). The barriers are one of the main reasons for a gap between consumer attitudes and behaviors. Based on the literature review, three main types of barriers have been identified:

1. *Product-based barriers* such as higher price of sustainable clothing, insufficient availability, poor attractiveness, usefulness and unattractive design, and so on.
2. *Communication-based barriers* related to the lack of information or insufficient information identifying sustainable clothing, which causes problems with distinguishing it from conventional products. These barriers arise from:

   - The complexity of information: an overwhelming number of data pointing to fair trade, organic cotton, use of animals, country of origin, and so on
   - The ambiguity of information: lack of specific/definitive information because of vague ethical policies, and the like
   - The incredibility of information
   - The distrust of the information source, poor reputation of clothing companies, and so on (Hassan et al. 2013)

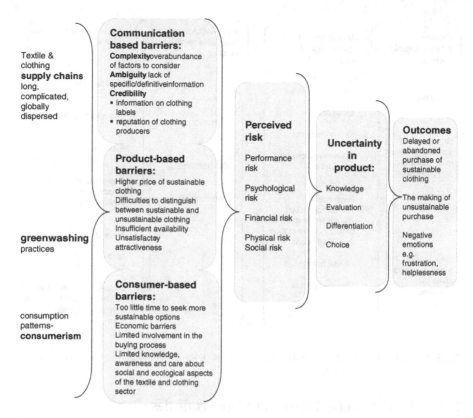

Fig. 6 Potential barriers discouraging the purchase of clothing. *Source* Developed by the author based on Hassan et al. (2013)

3. *Consumer-based barriers* arising from the lack of time to find more sustainable options, economic barriers, limited involvement in the buying process, limited knowledge, awareness and care about social and ecological problems in the textile and clothing sector, low receptiveness to communication.

All three categories of barriers make consumers feel exposed and uncertain. In the studies investigating the perceived risk effect on the intention to buy sustainable clothing, the following types of risks have been considered (Kang and Kim 2013; Han and Chung 2014):

- *Performance risk* arising from concerns about the function of a product; in the case of apparel the risk is higher because of insufficient information and confidence in product characteristics such as wearability and the ease of care, among others.
- *Psychological risk* related to concerns about a product being aesthetically inconsistent with the buyer's self-image. In the case of products made of organic cotton, there is also the aspect of limited selection and their being perceived as colorless and shapeless.

- *Financial risk* involved in the price of a product and the likelihood of financial loss. Sustainable clothing is usually priced higher than conventional items, because of very strict environmental, social, and ethical criteria the clothing has to meet. Additionally, its producers must supervise long supply chains. Most consumers fail to perceive higher prices of sustainable clothing as measurable benefits for their own and public health, and for the natural environment.
- *Physical risk* is related to sustainable clothing being potentially harmful to the user's health or being less appealing than the consumer expected. This risk may be unrealized when the consumer has insufficient knowledge of health hazards involved in unsustainable clothing products.
- *Social risk* related to the possibility of the product causing embarrassment or disapproval from one's family or peers.

The barriers to sustainable apparel consumption are presented in Fig. 6; they were also included in the theoretical model (Fig. 3).

6 Consumer Behavior Towards Sustainable Clothing—The Demand Side of the Model

This section is centered on customers and their behavior. Consumer behavior towards sustainable clothing should be understood as all customer's actions and perceptions involved in choosing a sustainable product, in decision making, and in finalizing the purchase.

In the first step, factors that significantly influence consumer behavior towards sustainable clothing and finally lead to a sustainable product being purchased are identified. Table 2 shows the results of earlier models investigating consumer behavior towards sustainable clothing.

As the models show, consumer behavior towards sustainable clothing is a very complex and multidimensional problem. The sections below provide a discussion of factors that earlier studies have pointed to as having significant influence on the purchase of sustainable clothing.

7 Environmental Knowledge, Awareness, and Concern

The first of the factors is consumer's knowledge of ecological and social threats involved in the production of clothing. This factor is given attention in many models (Kang et al. 2013; Moisander et al. 2010; Schlegelmilch et al. 1996; Thøgersen 2000; Young et al. 2010). Some authors argue, though, that knowledge alone is not sufficient (Brosdahl and Carpenter 2010). Much importance is also attached to the ecological and social awareness of the customer, understood as the ability to rationally engage one's knowledge resources in performing certain actions. One

Table 2 The main results yielded by models of consumer behavior towards sustainable clothing

Authors	Model constructs	Main outcomes
Shaw et al. (2007)	Attitude	Intention to avoid sweatshop apparel fully mediates the effects of its antecedents (attitude, subjective norm, perceived behavioral control, and desire) on the plan to avoid sweatshop apparel
	Subjective norm	
	Perceived behavioral control	
	Desire	
	Intention	
	Plan	
Halepete et al. (2009a)	Need for self-uniqueness	Attitude toward personalization of apparel is positively related to the intention to purchase personalized fair trade apparel
	Apparel involvement	
	Perceived financial risk toward buying apparel	
	Perceived social risk toward buying apparel	Consumers with greater need for self-uniqueness have positive attitude toward personalization and are unconcerned with social and financial risks.
	Body size—BMI	Body size and one's attitude toward personalized apparel are positively related to each other.
	Attitude toward personalization of apparel	
	Intention to purchase personalized fair trade apparel	
Brosdahl and Carpenter (2010)	Knowledge of environmental impacts	Knowledge of the environmental impacts of textile and apparel production leads to concern for the environment, which translates into environmentally friendly consumption behavior
	Concern for the environment	
	Environmentally friendly consumption behavior	
Ha-Brookshire and Norum (2011)	Product evaluative criteria	Consumer attitudes toward socially responsible apparel, attitudes toward environment, age, and gender are significant factors for consumers' willingness to pay a premium
	Demographic characteristics	Four apparel product evaluative criteria, namely brand name, laundering requirements, color, and fit are also important for consumers' willingness to pay a premium
	Apparel product evaluative criterion	
	Willingness to pay for socially responsible products	

(continued)

Table 2 (continued)

Authors	Model constructs	Main outcomes
Chan and Wong (2012)	Product-related attributes of eco-fashion (design, quality, price)	Store-related attributes of eco-fashion positively influence consumers' eco-fashion consumption decision, but this relationship can be weakened by the price premium level of eco-fashion.
	Store-related attributes of eco-fashion (customer service, store design and environment, store's environmental practices, shopping convenience)	
	Price premium level of eco-fashion	
	Eco-fashion consumption decision	
Hyllegard et al. (2012)	Subjective norm	Hang tags featuring highly explicit messages and third-party SR logos produce more favorable evaluations than do hang tags featuring less explicit messages and no logos.
	Gender	
	Clothing involvement	Consumers' clothing involvement as well as their past SR apparel purchasing behaviors predict their evaluations of apparel hang tags, which positively predicts their attitudes toward Good Clothes.
	Socially responsible apparel purchasing behaviors in the past	Attitude, subjective norm, clothing involvement, and past SR apparel purchasing behaviors predict patronage intention toward Good Clothes
	Attitude toward beliefs about "Good Clothes" engagement in sustainable production (SR) practices	
	Behavioral intention—intent to patronize the apparel brand "Good Clothes"	
Lee et al. (2012)	Perception of Green Product Brand	Perception of green product brands has positive impact on consumers' green behavior.
	Perception of Green campaign	Perception of green campaigns has a significant influence on consumers' green consciousness and an indirect impact on consumers' green behavior.

(continued)

Table 2 (continued)

Authors	Model constructs	Main outcomes
	Communication involvement	The relationship between consumers' perception of green campaign and green consciousness is stronger in the marketing communication involvement group
	Green consciousness	
	Green behavior intention	
Hassan et al. (2013)	Antecedents (complexity, ambiguity, conflict, credibility, source)	The main antecedents influencing uncertainties in decision-making processes about ethical clothing and resulting in delayed ethical purchase, the making of unethical purchase or negative emotions include:
	Uncertainty (knowledge, choice, evaluation)	Complexity: overabundance of factors to consider, for example, fair trade, organic cotton, use of animals, country of origin
	Outcomes (delayed purchase, compromised beliefs, negative emotions)	Ambiguity: lack of specific/definitive Information, for example, vague ethical policies
		Conflict: trade-off between products, attributes, and beliefs, for example, trade with poorer countries versus home-grown retailers
		Credibility: credibility of information on clothing labels
		Source: For example, the reputation of clothing organizations
Kang et al. (2013)	Consumer knowledge: consumers' familiarity with a product and product-specific knowledge	The study extended the TPB model by incorporating significant determinants of sustainable consumer behavior:
	Perceived consumer effectiveness	Consumer knowledge
	Perceived personal relevance: belief that the consumption of sustainable product is associated with one's personal lifestyle, value and self-image	Perceived consumer effectiveness
	Attitude: positive attitude to purchasing	Perceived personal relevance

(continued)

Table 2 (continued)

Authors	Model constructs	Main outcomes
	Subjective norm: perceptions about social pressure	The results indicate that those three determinants significantly affect young consumers' attitudes, subjective norms, and perceived behavioral control, thereby affecting purchase intentions for environmentally sustainable textiles and apparel
	Behavioral control: perceptions about the difficulty of the behavior	
	Behavioral intention	
Kang and Kim (2013)	Financial risk	The results have revealed the negative effect of perceived risk on consumers' attitudes and consequently on their intentions for buying sustainable textiles and apparel
	Performance risk	The significance and influence of perceived risks differ depending on the risk
	Psychological risk: consumer's concern with self-image	The greatest barrier keeping consumers from buying sustainable textiles and clothing is financial risk
	Social risk	Psychological risk directly and significantly shapes attitudes towards environmentally sustainable textiles and apparel consumption and indirectly shapes intentions for buying such products
	Attitude	The importance of social risk is smaller than thatof other risks
	Behavioral purchase intentions for environmentally sustainable textiles and apparel	The effect of performance risk on attitudes and behavioral intentions is not significant
Han and Chung (2014)	Perceived benefits of purchasing organic cotton apparel (environmental and health-related)	Perceived benefits, importance of individual expression through dressing well, performance risk, financial risk, and subjective norm significantly influence the attitude toward purchasing organic cotton apparel products
	Importance of individual expression through dressing well: subcategory of fashion orientation	Additionally, financial risk, attitude, and subjective norms significantly influence consumers' purchase intention

(continued)

Table 2 (continued)

Authors	Model constructs	Main outcomes
	Performance risk: concerns about the expected function of a product	Subjective norms have been found to play a critical role in the purchase process: subjective norms are one of the strongest antecedents of attitude among the six variables under consideration and exert a similar influence on purchase intention as attitude does
	Psychological risk: concern that a purchased product will clash with a consumer's self-image	
	Financial risk: concerns about the price of the product and possible financial loss	
	Attitude towards purchasing organic cotton apparel	
	Subjective norm	
	Purchase intention	

Source Developed by the author

of the main reasons why customers are not interested in sustainable clothing seems to be unawareness of the problems and risks related to conventional products that frequently contain carcinogenic substances, and so on, or of the social and/or ecological implications of fast fashion (Pookulangara and Shephard 2013). For sustainable products to be sought, knowledge, the awareness arising from it, and the individual's predisposition to be concerned must come together (Balderjahn et al. 2013; Brosdahl and Carpenter 2010; Butler and Francis 1997; Kim and Damhorst 1998; Pino et al. 2012). Concern is what determines consumer attitudes and motivations, and makes them seek and acquire sustainable clothing (Fig. 7).

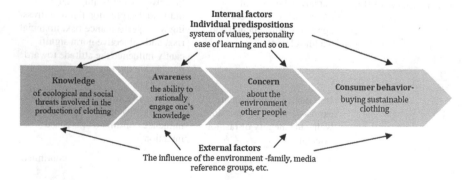

Fig. 7 Influence of environmental knowledge, awareness, and concern on consumer behavior. *Source* Based on Koszewska (2012b)

8 Perceived Consumer Effectiveness (PCE)

Perceived consumer effectiveness (PCE) is believed to be crucial to the purchase of products with ecological and/or social attributes. The concept of PCE was described for the first time by Kinnear, Taylor, and Ahmed (Kinnear et al. 1974), who presented it as a measure of a person's belief in his ability to contribute to the abatement of pollution effectively (Booi-Chen 2011). Roberts (1996) has defined the PCE as a measure of an individual's self-assessed ability to help solve environmental resource problems, finding it to be one of the most salient factors in explaining environmentally conscious consumer behavior. Mohr, Webb, and Harris (Mohr et al. 2001) have confirmed PCE's significant contribution to socially responsible consumer behavior and have demonstrated that the more consumers perceive their purchasing power as influencing company behavior, the more inclined they are to practice socially responsible behavior. Another study (Paço et al. 2013) has shown that consumers who have a low rating of their PCE are reluctant to spend their time or effort to shift from buying conventional products to Green products, regardless of whether they are environmentally concerned. All the authors concluded that consumers needed to feel empowered to be personally effective in combating environmental problems.

In the context of sustainable clothing, the PCE can be understood as the strength of consumers' belief that their decisions may induce clothing companies into ecological and pro-social behavior. The PCE influence on the intention to buy environmentally sustainable apparel has been analyzed in the Kang et al. model (Kang et al. 2013). Their model has shown that consumers believing in being able to improve the environment through careful consumption develop positive attitudes towards environmentally sustainable apparel, which indirectly increases the likelihood of such products being actually purchased.

9 Factors Indicated by the Theory of Attitude–Behavior Relationships

Among the most popular models explaining the attitude–behavior relationship there are the theory of reasoned action (TRA; Ajzen and Fishbein 1980) that links attitudes, subjective norms, behavioral intentions, and behavior into a fixed causal sequence, and the theory of planned behavior (TPB) (Ajzen 1991) that emerged after the original theory of reasoned action was extended to perceived behavioral control to account for behaviors that individuals cannot fully control (Shaw et al. 2007). These classic theories have been extensively adopted with a view to explaining consumer behavior towards sustainable products, including clothing. By analyzing models that have developed from them, other key factors influencing the purchase of sustainable clothing can be identified (Fig. 8).

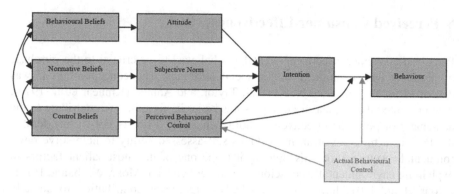

Fig. 8 Theory of planned behavior. *Source* Bray (2008)

9.1 Intentions

The Fishbein and Ajzen models are built around intentions that encompass all motivations guiding consumers and indicate the degree to which a consumer may be willing to show some behavior. According to the main assumption of these models, consumer behavior and consumer intentions are consistent unless special circumstances arise.

The majority of models explaining consumer behavior towards sustainable clothing treat intentions as an endogenous variable. The examples of such models are the model of the intention to purchase personalized fair trade apparel (Halepete et al. 2009a), the model of the behavioral intention to patronize the apparel brand "Good Clothes" (Hyllegard et al. 2012), the model of the Green behavior intention (Lee et al. 2012), the model of purchase intentions for environmentally sustainable textiles and apparel (Kang et al. 2013), and the model of organic cotton apparel purchase intention (Han and Chung 2014).

Among intentions, the willingness to pay more for sustainable clothing is very important. This variable has been explained by Ha-Brookshir and Norum, who have found the willingness to pay a premium for socially responsible apparel to be influenced by consumer attitudes towards this type of apparel and their milieu, age, and gender (Ha-Brookshire and Norum 2011).

Similarly, in the study by Hustvedt and Bernard the consumers' willingness to pay for apparel products displaying labor-related information and brand was the most strongly influenced by the sense of social responsibility and the attitude to fair trade (Hustvedt and Bernard 2010). Consumers appreciating social responsibility and fair trade were more inclined to pay a premium than those who were not. On the other hand, concerns about the environmental impact of apparel reduced the amount participants would bid for the labeled apparel. Hustvedt and Bernard's study has also showed that consumers would pay more for apparel with information about its labor-related attributes if they had the economic incentive to express their true value for the products. Another factor influencing the willingness to pay a premium was ethnicity (Hispanic participants bid higher when the information about labor conditions was attached to the products).

Because all these studies agree that the willingness to pay a premium is an important intention leading to a purchase of sustainable clothing, this variable, too, was included in the empirical submodel presented in Sect. 4.

9.2 Attitudes

Consumer behavior towards sustainable clothing is also significantly influenced by the attitude towards this behavior. According to Ajzen's theory, attitudes determined by person's beliefs about the expected outcomes of some behavior are the main component of each intention. The more positive attitude one has towards some behavior, the stronger the intention is to show it. The analyzed models of consumer behaviors towards sustainable clothing confirm that this relationship does exist. The attitude toward sweatshop avoidance significantly impacts the dependent construct, the intention to avoid sweatshop apparel (Shaw et al. 2007); the attitude towards personalization of apparel is positively related to the intention to purchase personalized fair trade apparel (Halepete et al. 2009a); a positive attitude toward "Good Clothes" influences patronage intention (Hyllegard et al. 2012) and, lastly, a positive attitude to purchasing environmentally sustainable textiles and apparel has a positive effect on purchase intentions (Kang et al. 2013).

9.3 Subjective Norms

Another group of factors that determine intentions according to TRA is subjective norms, which are closely related to the social pressure on showing some behavior or abstaining therefrom. Subjective norms are based on the person's conviction that some behavior will or will not be accepted by individuals whose opinion is valued by the person. Subjective norms are a particularly important determinant of human behavior towards clothing, because social pressures are frequently decisive for what people choose to buy. The influence of subjective norms on intentions and/or on the purchase of sustainable clothing has been confirmed empirically. The norms have been found to predict patronage intention toward "Good Clothes" (Hyllegard et al. 2012), to affect purchase intentions for environmentally sustainable textiles and apparel (Kang et al. 2013), and to influence a plan to avoid sweatshop apparel (Shaw et al. 2007). Subjective norms also play a critical role in the purchase of organic cotton apparel.

9.4 Perceived Behavioral Control

The last group of factors significantly affecting consumer intentions is the means and possibilities enabling a person to behave in a certain manner, that is, to buy sustainable clothing in our case.

This variable has already been used in the models of consumer behavior towards sustainable clothing. The results of studies into the influence of perceived behavioral control on buying intentions are inconsistent. Kang et al. (2013) have not found perceived behavioral control to have a major effect on the intention to buy environmentally sustainable textiles and apparel, but another study has demonstrated that perceived behavioral control significantly determines the intention to avoid sweatshop apparel (Shaw et al. 2007).

Consumer behavior towards sustainable clothing is also dependent on the individual's openness to communication.

10 Openness to Company's CSR Communication—Recognizability of Eco-Labels

Environmental labeling schemes are potentially effective and consumer-friendly tools for communicating the attributes of sustainable clothing, but they are not useful unless a consumer notices them in a shopping situation. Thogersen (2000) argues that seeing an eco-label is not a goal in itself, but rather a step towards buying a sustainable product. According to this author, an individual's ability to take note of an eco-label depends on his pro-environmental attitude and PCE. The important questions that must be asked here concern factors making eco-labels recognizable to consumers and the influence of the level of their recognizability on consumer buying decisions. To answer these questions, the empirical submodel proposed in Sect. 4 is provided with the variable "recognizability of eco-labels."

Another variable that the empirical models of consumer behavior towards sustainable clothing have not thoroughly analyzed thus far is shopping habits and apparel selection criteria.

11 Shopping Habits and Apparel Selection Criteria

The way consumers behave in the clothing market depends also on their shopping habits and the criteria they use in choosing products. These factors frequently stem from past experiences, the loyalty to a particular store or brand, store patronage, interest in fashion, and shopping orientation, as well as from the attitude to quality and aesthetics. All of them have been considered in the general models of consumer behavior towards textiles and clothing (De Klerk and Lubbe 2008; Du Preez

2003; Du Preez and Visser 2003) and partly in the models of consumer behavior towards sustainable apparel (Halepete et al. 2009b). The results of earlier studies show that consumers with greater apparel involvement (feeling a greater urge to be unique) have a positive attitude towards purchasing personalized fair trade apparel (Halepete et al. 2009b). Moreover, four apparel product evaluative criteria, namely brand name, laundering requirements, color, and fit proved important for the consumer's willingness to pay a premium for socially responsible apparel (Ha-Brookshire and Norum 2011). This crucial importance of shopping habits and apparel selection criteria for consumer behavior towards sustainable clothing has been confirmed in earlier studies (Hustvedt and Bernard 2010; Koszewska 2013). In Koszewska's study, they determined consumers' openness to CSR messages, preference for sustainable innovative textiles and clothing, motivations for buying a product, the willingness to pay a premium for sustainable clothing, as well as actual purchases thereof.

Chan and Wong have investigated the influence of product-related attributes and store-related attributes of eco-fashion on eco-fashion consumption decisions (Chan and Wong 2012). They have found that the product-related attributes such as design, quality, and price did not affect these decisions and that the only factor encouraging consumers to choose eco-fashion was store-related attributes such as customer service, store design and environment, store's environmental practices, and shopping convenience. All this points to shopping habits and apparel selection criteria as very important predictors of the purchase of sustainable apparel. However, because of inconsistent findings of the existing studies further analysis of these factors is necessary. For this reason, they were entered into the empirical submodel in Sect. 4.

The last group of factors analyzed in the context of sustainable clothing is of sociodemographic nature.

12 Sociodemographic Factors

The influence of the sociodemographic factors on sustainable product consumption has been addressed in many studies (Arcury and Johnson 1987, Cottrell and Graefe 1997; Diamantopoulos et al. 2003; Jain and Kaur 2006; Mohai and Twight 1987). Most of them focus on green consumer behavior, particularly on purchasing and disposal. Their conclusions regarding the impacts of sociodemographic factors are often divergent and contradictory (Auger et al. 2008; Jain and Kaur 2006; McDonald and Oates 2006).

12.1 Education

An individual's education significantly contributes to the development of environmental attitudes, because a substantial amount of environmental knowledge that comes with education is required to properly understand environmental problems (Arcury 1990; Liu et al. 2012). Many studies point to a positive correlation between education and environmental knowledge (Arbuthnot and Lingg 1975; Arcury and Johnson 1987; Diamantopoulos et al. 2003; Maloney and Ward 1973; Ostman and Parker 1987), environmental attitudes (Baker and Bagozzi 1982; Leonard-Barton 1981; Roberts 1996; Tognacci et al. 1972; Dunlap and Van Liere 1981; Zimmer et al. 1994), and environmental behaviors (Arbuthnot 1977; Harry et al. 1969; Jolibert and Baumgartner 1981; Ostman and Parker 1987; Scott and Willits 1994, Webster Jr 1975). In some studies, however, the correlation between education and environmental attitudes and/or behavior was found to be negative (Arbuthnot and Lingg 1975; Samdahl and Robertson 1989), or no significant correlation between education and environmental knowledge, attitudes, or even behavior was established (Diamantopoulos et al. 2003; Honnold 1981; Neuman 1986; Pickett et al. 1993; Ray 1975; Shrum et al. 1995). An interesting case is Jain and Kaur's study (Jain and Kaur 2006), where the level of respondents' education was significantly but negatively correlated with their willingness to seek out eco-friendly products.

12.2 Income

In most empirical studies investigating the relationship between income and the theoretical dimensions of the environmental consciousness domain such as knowledge, attitudes, and behavior, the factors were correlated significantly and positively (Arbuthnot 1977; Chan 2000; Kinnear et al. 1974; Tai and Tam 1996; Tucker and Dolich 1981; Webster Jr 1975; Zimmer et al. 1994). Several empirical studies have shown, though, that the relationship between income and environmental behavior is not significant or even negative (Antil 1984; Buttel and Flinn 1978; Kassarjian 1971; Pickett et al. 1993; Roberts 1996; Samdahl and Robertson 1989; Dunlap and Van Liere 1981). It is, therefore, another area where more research is necessary to establish whether a person's financial capacity may directly determine the purchase of sustainable clothing.

Figure 9 explains the demand side of the proposed theoretical model more in detail, taking account of all factors that the literature considers to influence significantly the purchase of sustainable clothing.

The empirical verification of all relationships between variables used in the theoretical model is not straightforward, because the relationships are much more complicated than the diagram shows. To make the process of verification easier to follow, in the next section only selected variables determining the purchase of

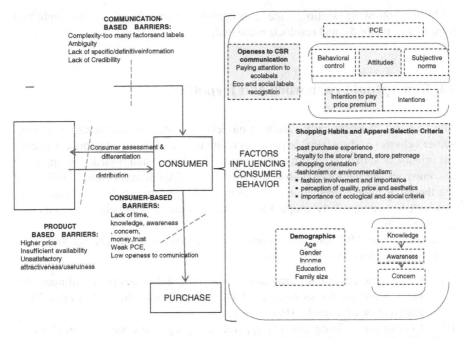

Fig. 9 Factors in purchasing sustainable clothing: a closer look into the demand side of the theoretical model

sustainable clothing undergo empirical verification. The variable selection criteria include the significance of a variable, few earlier studies, or inconsistent results of earlier studies. In Fig. 9, the selected variables are highlighted in grey.

13 The Empirical Verification of the Demand Side of the Model

The empirical submodel concentrates on several factors important for consumer behavior towards sustainable clothing, namely:

- Shopping habits and apparel selection criteria understood as attitudes to apparel shopping
- Willingness to pay a premium for sustainable clothing
- Openness to communication understood as the receptiveness of ecological and social labels on clothes
- Purchase of sustainable clothing

A detailed discussion of the significance of the selected variables and their theoretical background has been provided in the section above. Section 4 focuses on their conceptualization and operationalization.

In the following sections, the conceptual framework of the research and hypotheses, methods, and results is presented.

14 Conceptual Framework and Hypotheses

The majority of empirically validated models described in the literature on consumer behavior and sustainable products relate to one of the most popular theoretical orientations, that is, cognitive orientation. As already mentioned, this approach encompasses the theory of reasoned action (TRA) (Ajzen and Fishbein 1980) and the theory of planned behavior (TPB) that developed from it. These classic theories have been extensively adopted to explain how consumers behave towards sustainable clothing.

Based on the TPB and conclusions from the discussed models, the following hypotheses are formulated (Fig. 10).

H1:   Consumers' attitude towards apparel shopping influences their willingness to pay a premium for sustainable clothing (Ha-Brookshire and Norum 2011; Hustvedt and Bernard 2010).

H2:   Consumers' attitude towards apparel shopping influences the purchase of sustainable clothing (Halepete et al. 2009b; Hustvedt and Bernard 2010; Koszewska 2013).

H3:   Consumers' attitude towards apparel shopping influences the recognizability of ecological and social labels (Thøgersen 2000).

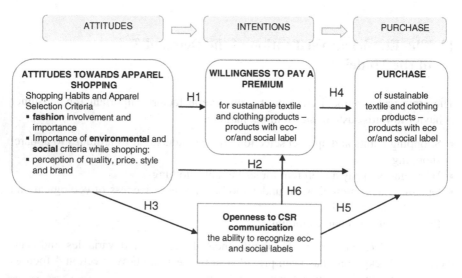

Fig. 10 The conceptual model

H4:   Consumers' willingness to pay a premium for sustainable clothing is positively correlated with the purchase of sustainable clothing (see the theory of planned behavior (TPB)).

An important variable that appears to determine the purchase of a sustainable product, but which has been insufficiently covered in the literature thus far, is the ability to recognize ecological and social labels attached to sustainable apparel. To verify empirically whether the recognizability of labels significantly determines the purchase of such products as well as the willingness to pay a premium for them, the following hypotheses are formulated in this study.

H5:   The recognizability of ecological and social labels positively influences the purchase of sustainable clothing.

H6:   The recognizability of ecological labels positively influences the willingness to pay a premium for sustainable clothing.

15 Methods

15.1 Survey and Sample

A survey of a sample of 981 Polish adults drawn from the official population database was conducted. To make sure that the sample was representative and the survey was reliable, the sampling procedure and the survey interviews were carried out by the Public Opinion Research Centre (CBOS). The interviews were conducted using the CAPI technique (computer-assisted personal interviewing). The sociodemographic characteristics of the respondents are presented in Table 3.

Table 3 Sociodemographic structure of Polish respondents

Gender	Male	47.6 %
	Female	52.4 %
Age	18–24 years	13.6 %
	25–34	17.4 %
	35–44	14.6 %
	45–54	18.1 %
	55–64	18.1 %
	65 years and older	18.3 %
Place of residence	Rural areas	37.6 %
	Town with population to 20,000	13.9 %
	20,000–100,000	20.0 %
	101,000–500,000	15.8 %
	501,000 and more	12.7 %

(continued)

Table 3 (continued)

Education	Elementary	25.3 %
	Basic vocational	25.7 %
	Secondary	33.7 %
	Tertiary	15.3 %
Socio-occupational group Economically active	Managers, specialists with tertiary education	17.4 %
	Middle-level personnel, technicians	8.2 %
	Office and administrative personnel	13.7 %
	Personnel in the service sector	10.6 %
	Skilled workers	23.2 %
	Unskilled workers	11.2 %
	Farmers	9.3 %
	Self-employed	6.5 %
Economically inactive	Disability pensioners	13.6 %
	Old-age pensioners	44.1 %
	School-children and students	14.0 %
	Unemployed	17.4 %
	Housewives and others	11.0 %

16 Methodology and Analytical Procedures

The theoretical model presented above was analyzed using the CFA (confirmatory factor analysis) and SEM methods in three steps. First, the exploratory factor analysis (EFA) and the principal components method with varimax rotation were used to establish the basic structure of the selected variables. In the second and third steps the model was tested with, respectively, the CFA and the SEM (Davis and Lang 2012).

17 Conceptualization and Operationalization of the Empirical Submodel's Variables

17.1 Shopping Habits and Apparel Selection Criteria

The earlier models analyzed the following constructs relating to shopping habits and apparel selection criteria.

- *The need for self-uniqueness* (to be different from others) that Halapete measured with an eight-item scale of uniqueness provided with a five-point Likert scale (Halapete et al. 2009b)
- *Involvement in shopping* for apparel established by asking the respondents to rate their general feelings while shopping for apparel on a five-point bipolar scale (important/unimportant, boring/interesting, etc.; Halapete et al. 2009b)

- *Apparel evaluative factors* (fit, price, laundering (or care), style, color, brand) ranked by respondents on a five-point scale (1: insignificant and 5: very important; Ha-Brookshire and Norum 2011)
- *Product-related attributes* product design, quality, price
- *Shop-related attributes* (customer service, store display and environment, store's ethical practices, and shopping convenience) ranked by respondents on a five-point scale (1: very unimportant and 5: very important (Chan and Wong 2012)

Most of these models disregarded the ecological and social criteria that are crucial to buying, or not buying, sustainable clothing. To make up for this deficiency, the range of typical reasons for which consumers select and purchase apparel (fashion, brand and style, price, quality, wearability, comfort of wear, fit, the shopping location, etc.) was extended in this study to ecological and social criteria. In order to establish consumers' attitudes to buying clothing, a special seven-point bipolar scale with 10 items such as attitude to fashion, brand, ecology, raw materials, and social issues such as child labor and working conditions (Table 4) was developed and tested in the empirical submodel.

The scale's reliability and validity were determined with the Cronbach's alpha reliability test. Its results confirmed that the scale was reliable and valid for the representative sample of Polish consumers used in the survey (Cronbach's alpha > 0.8).

Table 4 Measures of attitudes towards apparel shopping

Specify which statements are closest to your habits. To make the assessment, use the 1–7 scale, where 1 means you are closest to the statement on the left, and 7 means that you are closest to the statement on the right

I'm not interested in fashion and new trends at all	1	2	3	4	5	6	7	I'm very much interested in fashion and new trends
I don't like original, unique, designer clothes at all	1	2	**3**	4	5	6	7	I like original, unique, designer clothes very much
The brand is completely unimportant to me	1	2	3	4	5	6	7	The brand is very important to me
I never buy clothes of global or European brands	1	2	3	4	5	6	7	I like to buy clothes of global or European brands very much
I never check for raw materials composition	1	2	3	4	5	6	7	I always check for raw materials composition
I never check for the producer country	1	2	3	4	5	6	7	I always check for the producer country
I never check if the clothes have eco-labels or eco-symbols	1	2	3	4	5	6	7	I always check if the clothes have eco-labels or eco-symbols
I never think of whether the product involves child labor	1	2	3	4	5	6	7	I frequently think of whether the product involves child labor
I never think of whether the rights of the workers making the product were infringed	1	2	3	4	5	6	7	I frequently think of whether the rights of the workers making the product were infringed
I do not pay any attention to whether or not the clothes are natural/organic	1	2	3	4	5	6	7	I only buy natural/organic clothes

18 Openness to Companies' CSR Communication— Recognizability of Ecological and Social Labels

Although consumers' openness to messages coming from the producers is one of the key factors driving the development of a sustainable clothing market, there are few empirical studies that have investigated this area. Thogersen analyzed factors influencing the variable "paying attention to ecolabels," finding it to be a major determinant of a buying decision. The variable was measured with the following question, "When you are choosing a product, how often do you pay attention to any environmental labelling before deciding to buy?" The possible answers were "often", "sometimes", or "never" (Thogersen 2000).

Ecological and environmental labels attached to apparel are easy to see and despite their shortcomings they are one of the best means of communicating a company's CSR. The recognizability of such labels was therefore introduced into the model as a measure of consumer's openness to CSR messages. The respondents were shown seven ecological and social labels known in Poland (EU's Eco label, Global Organic Textile Standard (GOTS) and Oeko-Tex Standard, Fair trade, Fair Wear Foundation (FWF), and two Polish labels (Infant Safe and Child Safe) and were asked to mark their answers to the question: "Labels on textile products such as clothing, bedclothes, towels and curtains are identified by different symbols. Have you seen this label?" on the attached scale (yes = 1; no = 0). Therefore, the variable "recognizability of ecological and social labels" could take values ranging from 0 to 7, depending on how many labels and signs the respondent recognized. Because of the dichotomously scored items, to evaluate the reliability of the scale the Kuder–Richardson reliability coefficient (KR20) was employed. The Kuder–Richardson statistics (KR20 = 0.75) pointed to high reliability of the scale.

19 Consumers' Willingness to Pay a Premium for Sustainable Clothing and to Buy Sustainable Clothing

To establish consumers' willingness to pay a premium for sustainable clothing, the respondents were asked to mark their answers to the question "What premium would you pay for an article of clothing if you believed it had been made using a process safe for human health and the environment and respecting the workers' rights?" on 5-point scale (to 5 % more; between 6 and 10 % more; between 11 and 15 % more; between 16 and 20 % more; over 20 %; I would not pay a premium).

The variable representing the actual purchase of sustainable clothing[4] was measured with the question "Have you ever bought apparel carrying an ecological

---

[4]One of the best sources of information on apparel sustainability available to consumers is ecological and social labels. In this study, sustainable apparel is meant as items with ecological and/ or social labels.

and/or ethical label?" and a 3-point scale (1 = no; 2 = I don't remember; 3 = yes). A variable measured with such as a scale cannot be treated as a continuous variable, but the scale was used notwithstanding because McEachern and Carrigan (2012) recommend analyzing what consumers actually buy and not what they plan or intend to buy when studying ethical consumption behavior in order to reduce the attitude–behavior gap (Carrigan and Attalla 2001; Auger et al. 2008). To prevent the "purchase" variable from causing methodological errors, a polychoric correlation matrix was used (Olsson 1979; Plesniak 2006).

20 Results

20.1 Exploratory Factor Analysis

A high value of the Kaiser–Meyer–Olkin (KMO) indicator (0.81) and the result of the Bartlett test supported the choice of a factor analysis model for data analysis.

The EFA results showed that a three-factor model would be appropriate. Ten observable variables presented in Table 5 were aggregated into three factors (attitudes) towards apparel shopping, namely "fashion", "ecology", and "ethics". The results of the EFA are shown in Table 5.

Factor 1 was made up of variables such as clothing brand names, originality, fashion trends, and design and was called "a pro-fashion attitude". Factor 2 was compiled of variables such as raw material composition, country of production,

Table 5 The structure of the principal components (Factor Loadings) after rotation for apparel selection criteria and shopping habits

Apparel selection criteria and shopping habits			
Observable variables	Factors (Principal Components)		
	Fashion	Ecology	Ethics
Buying clothes of global or European origin	0.842		
Preference for original, unique, designer clothes	0.833		
Brand importance	0.765		
Interest in fashion and new trends	0.695		
Assessment of raw material composition		0.807	
Preference for clothes from natural raw materials		0.783	
Checking for producer country		0.779	
Checking for eco-labels		0.607	
Thinking of workers' rights when buying clothes			0.904
Thinking of child labor when buying clothes			0.900
Percentage of variance explained	**38.4**	**19.2**	**11.0**
Cronbach's alpha	**0.82**	**0.8**	**0.86**

Source Calculated by the author using IBM SPSS software

and ecological labeling. All these variables are related to ecology, so factor 2 was named "a pro-ecological attitude". Factor 3 was constructed using variables such as possible violations of workers' rights and the use of child labor. All of them concern social and ethical issues, so factor 3 was called "a pro-social attitude".

21 Confirmatory Factor Analysis: The Measurement Model

To verify the validity of the three factors (attitudes), a confirmatory factor analysis was conducted with SPSS/AMOS 20.0. Convergent validity was determined by means of three different measures: factor loadings, average variance extracted (AVE), and construct reliability (CR; Baumgartner and Homburg, 1996). The CFA results suggested that two items should be removed (because of factor loadings <0.6). The final model had a satisfactory level of validity. The values of average variance extracted (AVE) were not below the threshold of 0.50 recommended for each construct. All CFA loadings ranged from 0.74 to 0.89, and the reliability of the construct used in testing internal consistency exceeded the recommended threshold of 0.7 (Fornell and Larcker 1981; Hair et al. 2011). Table 6 presents the relevant correlations, AVE values, covariances, and the construct reliability for all latent variables in the model.

The goodness-of-fit indices obtained from the measurement model are good. The model's normed $\chi 2$ was 3.4, the root mean square error of approximation (RMSEA) = 0.05, the goodness-of-fit index (GFI) = 0.99, adjusted GFI (AGFI) = 0.97, and the comparative fit index (CFI) = 0.98 (see Table 6). Accordingly, the measurement model was accepted for analysis.

22 Structural Model and Hypothesis Testing Results

After a reasonably well-fitting measurement model was identified, the structural model was tested by means of structural equation modeling (SEM). Figure 11 shows the final version of the empirical submodel with standardized estimates of structural paths and fit indices. The proposed structural model has a very good fit, because the values of all fit indices exceed the recommended values: $\chi 2$ (df = 33) = 75.37, p = 0.00; GFI = 0.98; AGFI = 0.96; CFI = 0.98; RMSEA = 0.043; and N Holter = 445.

The results of hypothesis testing are shown in Table 7.

The results yielded by the structural model support all hypotheses. Hypotheses H1a, H1b, and H4b state that pro-fashion, pro-environmental, and pro-social attitudes strengthen consumers' willingness to pay a premium for sustainable apparel. The results of statistical analysis show that all three of them are important, thus hypothesis H1 is fully confirmed.

Table 6 The CFA results

Constructs—latent variables	Indicator	Measurement model reliability and validity				
		CFA loading*	AVE**	CR***	Cronbach's Alpha
Pro-fashion attitude	Buying clothes with global or European brands	0.75	0.6	0.79	0.82
	Preference for original, unique, designer clothes	0.75			
	Brand importance	0.84			
Pro-ecology attitude	Checking for raw material composition	0.78	0.54	0.78	0.76
	Checking for producer country	0.74			
	Preference for clothes from natural raw materials	0.7			
Pro-social attitude	Thinking of workers' rights when buying clothes	0.89	0.75	0.86	0.86
	Thinking of child labor when buying clothes	0.84			

Measurement model fit	
$\chi 2 = 57$, (df 17)	RMSEA < 0.05–0.08 (Browne and Cudeck 1992)
$\chi 2/df = 3.4$	GFI > 0.90 (Bollen 1989; Marcoulides and Schumacker 1996)
RMSEA = 0.05	AGFI > 0.90 (Bollen 1989; Marcoulides and Schumacker 1996)
GFI = 0.99	CFI > 0.90 (Bollen 1989; Marcoulides and Schumacker 1996)
AGFI = 0.97	
CFI = 0.98	
IFI = 0.98	
N Holter = 462	

*CFA loading: standardized estimate

**Average variance extracted (AVE) $= \sum$ (standardized loadings2)/$\sum$ (standardized loadings2) $+ \sum$ error variance

***Construct reliability (CR) $= \sum$ (standardized loadings) 2$\sum$ (standardized loadings) 2 $+ \sum$ error variance

Source Calculated by the author with the IBM SPSS/AMOS software

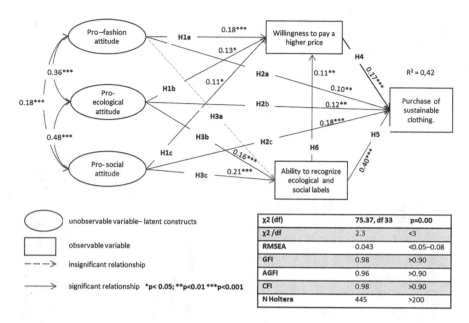

Fig. 11 The tested model and SEM results: the results of hypothesis testing. *Note* All are standardized estimates

Hypothesis H2 predicting that the consumers' attitude towards apparel shopping and sustainable clothing purchase are related to each other has been proven too. According to statistical analysis, pro-fashion, pro-ecological, and pro-social attitudes are significant predictors of purchasing sustainable clothing, which confirms hypotheses H2a, H2b, and H2c.

Hypothesis H3 states that the consumers' attitude towards apparel shopping influences their ability to recognize ecological and social labels. Given that the pro-ecological and pro-social attitudes were found to be positively correlated with this ability, hypotheses H3a and H3b are proven true as well. However, for the lack of significant relation between a pro-fashion attitude and the recognizability of eco-labels H3a must be rejected. Hypothesis H3 is therefore only partially confirmed.

According to the results of statistical analysis and the TPB-based predictions, consumers' willingness to pay a premium for sustainable clothing is positively related to the purchase of such products, therefore hypothesis H4 is true.

The analysis has also pointed out that customers with a greater ability to recognize ecological and social labels attached to apparel are more likely to purchase sustainable clothing ($\beta = 0.40$, $p < 0.001$), which supports hypothesis H5. Moreover, there is a significant path coefficient between the recognizability of ecological labels and an individual's willingness to pay a premium for such clothing ($\beta = 0.11$, $p < 0.01$). These consumers are more willing to pay a higher price for sustainable clothing, thus hypothesis H6 is proven true.

Table 7 Regression results for the hypotheses

HP	Path	β	p	Outcome
H1	Consumers' attitude towards apparel shopping influences their willingness to pay a premium for sustainable clothing			Fully supported
H1a	Pro-fashion attitude → willingness to pay a premium	0.17	$p < 0.001$	Supported
H1b	Pro-ecological attitude → willingness to pay a premium	0.13	$p < 0.05$	Supported
H1c	Pro-social attitude → willingness to pay a premium	0.11	$p < 0.05$	Supported
H2	Consumers' attitude towards apparel shopping influences sustainable clothing purchase			Fully supported
H2a	Pro-fashion attitude → purchase of sustainable clothing	0.10	$p < 0.01$	Supported
H2b	Pro-ecological attitude → purchase of sustainable clothing	0.12	$p < 0.01$	Supported
H2c	Pro-social attitude → purchase of sustainable clothing	0.11	$p < 0.001$	Supported
H3	Consumers' attitude towards apparel shopping influences the recognizability of ecological and social labels			Partially supported
H3a	Pro-fashion attitude → labels' recognizability	0.04	$p = 0.34$	Not supported
H3b	Pro-ecological attitude → labels' recognizability	0.17	$p < 0.001$	Supported
H3c	Pro-social attitude → labels' recognizability	0.21	$p < 0.001$	Supported
H4	Willingness to pay a premium → purchase of sustainable clothing	0.17	$p < 0.001$	Supported
H5	Recognizability of ecological labels → purchase of sustainable clothing	0.40	$p < 0.001$	Supported
H6	Recognizability of ecological labels → willingness to pay a premium for sustainable clothing	0.11	$p < 0.01$	Supported

Note β = standardized regression weights

The squared multiple correlation of sustainable clothing purchase was 0.42, meaning that 42 % of purchase variance is explained by the predictor variables. Considering that the model was built with selected determinants of the purchase of sustainable clothing, this value seems to be satisfying.

23 Discussion

The aim of this study was to expand the knowledge of consumer behavior in the sustainable apparel market. In its empirical part, Polish consumers' intentions and attitudes to buying sustainable clothing were examined using the modified TPB (Ajzen and Fishbein 1980) and the existing models of consumer behavior towards sustainable apparel products.

In the course of analysis, three types of consumers' attitudes were derived from their shopping habits and apparel selection criteria: pro-fashion, pro-ecology, and pro-social. The focus of the study was not only on the significance of these attitudes, but also on the strength of their influence on the analyzed dimensions of consumer behavior, that is, the willingness to pay a premium for sustainable clothing, the recognizability of labels, and actual purchases of sustainable clothing. The earlier studies' conclusions about pro-environmental and pro-social attitudes positively influencing consumer's willingness to pay a premium (Halepete et al. 2009b; Ha-Brookshire and Norum 2011) and their ability to recognize an ecolabel (Thogersen 2000) have been supported by this research.

All three studies suggest that fashion may significantly determine the decision to buy sustainable apparel, but the relationship between these two variables has not been explicitly confirmed. Therefore, establishing whether consumers' attitudes towards fashion positively influence the analyzed aspects of their behavior seems as interesting as important.

The findings of this study imply that a pro-fashion attitude is the strongest predictor of one's willingness to pay a premium for sustainable clothing ($\beta = 0.17$ at $p < 0.001$ compared with 0.13 at $p < 0.05$ for a pro-ecological attitude and 0.11 at $p < 0.05$ for a pro-social attitude). This suggests that consumers whose shopping decisions are guided by a pro-fashion attitude (i.e., giving priority to original designer clothes, sensitive to brands, tending to choose clothes of global or European brands) are also more willing to pay a premium for sustainable clothing, probably because of the ease with which pro-fashion consumers join in the trend for eco-fashion. Higher prices of sustainable clothing do not discourage them from buying such items, because the brand and fashion are more important for them than price. The study has also shown that a pro-fashion attitude does not significantly contribute to the recognizability of ecological and social labels marking sustainable apparel, unlike pro-ecological and pro-social attitudes. It is probably due to the pro-fashion consumers' relatively careless attitude to buying sustainable clothing. They buy it because they tend to follow fashion and trends, whereas the ability to recognize an eco-label requires some involvement, awareness, and sensitivity. Consequently, the most important determinants of the consumer's ability to identify a label is a pro-social attitude (the standardized path coefficient value for this relationship is the highest for this relationship).

Nevertheless, all three shopping attitudes similarly and directly contribute to the purchase of sustainable clothing ($\beta = 0.10$, $p < 0.01$ for the pro-fashion attitude, 0.12, $p < 0.01$ for the pro-ecological attitude, and 0.11, $p < 0.001$ for the pro-social attitude). This result is consistent with the findings of earlier studies (Koszewska 2013) pointing to shopping habits and apparel selection criteria as the determinants of the purchase of sustainable clothing, but it is somewhat different from that of Moon et al. (2014). They did not find any significant correlation between fashion involvement and Green product consumption. However, one has to be cautious comparing the results of the two studies for several reasons: first, because of the size of the samples (327 vs. 981 respondents) and second, because of cultural differences between people in Asia and Central and East Europe (Polish vs. South Korea consumers), and third, because of different approaches to conceptualizing and operationalizing the scale for measuring the pro-fashion attitude (a 3-item scale focused on the preference for original designer clothes, global or European brands, and brand sensitivity versus a 10-item scale measuring involvement in fashion).

Earlier studies into consumers' willingness to pay a premium for sustainable apparel focused on its determinants, but omitted its influence on actual purchases. This study addressed this gap. Its results indicate that the willingness to pay a higher price for sustainable apparel is positively correlated with the purchase of sustainable apparel, which is consistent with the TPB intention–purchase relation (Ajzen and Fishbein 1980).

The certification and labeling systems are thought to have a major role in driving the expansion of sustainable clothing markets (Dickson 2001; Hartlieb and Jones 2009; Thøgersen 2000). Being one of the most effective and consumer-friendly channels for communicating the properties of sustainable clothing (Koszewska 2011a), the systems are not useful unless a consumer notices and recognizes a label or a symbol. In contrast with earlier studies that only dealt with factors influencing consumer's ability to recognize an eco-label, this study additionally sought to establish whether, and to what extent, the ability to recognize an eco- and/or social label influences the willingness to pay more for sustainable apparel and to purchase it. Its findings imply that the recognizability of ecological and social labels is one of the most important predictors of purchasing sustainable apparel (the standardized path coefficient value for this relationship is the highest among all purchase predictors $\beta = 0.40$, $p < 0.001$). The findings are also empirical evidence that the recognizability of labels has a major effect on consumers' willingness to pay more for sustainable apparel. All this confirms that labeling systems that are transparent and understandable to consumers play a key role in the expansion of sustainable clothing markets.

24 Conclusions

The aim of the theoretical and experimental research presented in the chapter was to expand the current knowledge of consumer behavior towards a specific group of products, sustainable apparel.

Despite numerous publications on consumer behavior modeling, a comprehensive model of consumer behavior with respect to sustainable clothing has not been created yet. The model proposed in the chapter is an attempt at completing this gap. The model reviews consumer behavior in the market for sustainable clothing and demonstrates the complexity of the relations between the socially responsible manufacturer of clothing and the consumer, as well as intermediate connections and influences between the key elements of the model:

- The producer and sustainable production systems
- Sustainable clothing product
- Barriers to sustainable consumption
- The consumer and factors determining her behavior

As indicated by the model, the following instruments are crucial to implementing sustainable production systems:

- Corporate social responsibility
- Eco/sustainable design as a means of improving product characteristics
- Life-cycle approach optimizing the production process
- Sustainable supply chain management
- Sustainability communication
- Slow fashion–slow production aiming to prolong the clothing's life cycle and maximize its utility and fuctionality, indicating slow consumption
- Eco-functional assessment

In the course of analysis, three causes of barriers to sustainable consumption were identified:

- Sustainable clothing and its undesirable features such as higher price, limited appeal, usefulness and design, and so on
- Ineffective communication: lack, complexity, and ambiguity of information, distrust of the information source, and so on
- Consumer characteristics: lack of time, knowledge and awareness, carelessness, limited receptiveness to communication, and so on

Complex theoretical models such as those proposed in the chapter are defined in the literature as large-system models. They enable a comprehensive depiction of the analyzed issue and help understand consumer behavior rather than predicting it. However, the complexity of the relations they try to capture and explain causes the empirical verification of these models to be very problematic (Bettman and Jones 1972). For this reason, only selected elements of the theoretical model proposed in this study, that is, consumers and the key determinants of their behavior towards sustainable clothing, were verified empirically. The conclusions from this part of research are the following.

1. Consumers' attitudes towards apparel shopping have a significant and positive influence on their willingness to pay a premium for sustainable clothing, the recognizability of ecological and social labels, and, finally, on the actual purchase of sustainable clothing. More specifically:

   - They positively influence the pro-ecological and pro-social attitudes on consumers' willingness to pay a premium for sustainable apparel and the recognizability of apparel labels.
   - A pro-fashion attitude may exert a relatively strong influence on the willingness to pay a premium for sustainable clothing (unlike pro-ecological and pro-social attitudes) but does not significantly affect the recognizability of ecological and social labels attached to it.

2. The majority of earlier studies did not analyze the purchase of sustainable apparel, but they focused on the intention to buy (which may not be realized). Therefore, this research makes a unique contribution to the body of knowledge on sustainable apparel consumption as it provides:

- Empirical evidence of the influence of consumers' attitudes (pro-fashion, pro-ecological, and pro-social) towards apparel shopping, of their willingness to pay a premium, and of the recognizability of apparel labels on actual purchases of sustainable apparel.
- Additional evidence pointing to a significant role of the recognizability of ecological and social labels in purchasing sustainable apparel. This recognizability not only directly contributes to the decision to buy a sustainable product, but also makes the consumer more willing to pay a premium for it.

The findings of this study are likely to be useful for both marketers and public policy makers wishing to promote a wider use of sustainable apparel and foster the expansion of markets for sustainable products. They show that different shopping attitudes can motivate consumers to buy sustainable apparel. Consumers with pro-ecological and pro-social attitudes, but also those with a pro-fashion attitude, may appreciate the added value of sustainable clothing and buy it, even if they have to pay a premium. However, different categories of consumers would do that for different reasons. Companies should be aware of the different characteristics of consumers when implementing long-term CSR strategies, as well as making marketing decisions on a current basis.

This research also highlights the critical role of the recognizability of ecological and social labels in purchasing sustainable clothing. Consumers must have all necessary data to make informed decisions. Clothing companies should therefore work on increasing their knowledge and awareness of products and labels used, and make their CSR communication systems more transparent and friendly to consumers.

25 Limitation and Future Research

The study has some limitations that may affect the general applicability of its findings. Most of all, the sample used consists of Polish consumers, so researchers should be cautious in generalizing the findings of this study to other countries, although comparing them with the results obtained for countries would be a very interesting exercise (e.g., with new consumer countries with slowly developing ethical consumer movements such as Poland and long-rich countries leading in ecological and ethical consumer movements). Another limitation is that the scales used for measuring the construct of pro-social attitude have two items instead of the recommended three and that the empirical verification involved a limited number of variables. Future research should address more factors that are indicated as having significant influence on the purchase of sustainable clothing.

This research focused on consumer behavior, but future investigations should also consider the behavior of the producers of sustainable clothing and the factors influencing it. The theoretical model that only addresses barriers to consumption should be extended in the future to account for barriers to sustainable production.

Acknowledgements The research was partially supported by Scientific grant awarded by the Rector of the Lodz University of Technology. I would like to thank Janusz Kwitecki for his work on the translation and Agnieszka Pleśniak from the Warsaw School of Economics for providing insight and expertise in the field of structural equation modeling. I would also like to express my gratitude to Prof. Zofia Wysokińska and anonymous reviewers for their valuable comments on an earlier version of the manuscript which helped me improve the final version of the chapter

References

Ajzen I (1991) Theories of cognitive self-regulation. The theory of planned behavior. Organ Behav Hum Decis Process 50:179–211

Ajzen I, Fishbein M (1980) Understanding attitudes and predicting social behavior. Prentice-Hall, Englewood Cliffs NJ

Antil JH (1984) Socially responsible consumers: profile and implications for public policy. J Macromarketing 4:18–39

Arbuthnot J (1977) The roles of attitudinal and personality variables in the prediction of environmental behavior and knowledge. Environ Behav 9:217–232

Arbuthnot J, Lingg S (1975) A comparison of French and American environmental behaviors, knowledge, and attitudes. Int J Psychol 10:275–281

Arcury T (1990) Environmental attitude and environmental knowledge. Hum Organ 49:300–304

Arcury TA, Johnson TP (1987) Public environmental knowledge: a statewide survey. J Environ Educ 18:31–37

Auger P, Devinney TM, Louviere JJ, Burke PF (2008) Do social product features have value to consumers? Int J Res Mark 25:183–191

Baker DA, Bagozzi RP (1982) Attitudes toward public policy alternatives to reduce air pollution. J Public Policy Mark 1:85–94

Balderjahn I, Peyer M, Paulssen M (2013) Consciousness for fair consumption: conceptualization, scale development and empirical validation. Int J Consum Stud 37:546–555

Baumgartner H, Homburg C (1996) Applications of structural equation modeling in marketing and consumer research: a review. Int J Res Mark 13:139–161

Bettman JR, Jones JM (1972) Formal models of consumer behavior: a conceptual overview. J Bus 45:544–562

Bhattacharya CB, Sen S (2004) Doing better at doing good: when, why, and how consumers respond to corporate social initiatives. Calif Manag Rev 47:9–24

Blowfield ME (1999) Ethical trade: a review of developments and issues. Third World Q 20:753–770

Bollen KA (1989) Structural equations with latent variables. Wiley, New York

Booi-Chen T (2011) The role of perceived consumer effectiveness on value-attitude-behaviour model in green buying behaviour context. Aust J Basic Appl Sci 5:1766–1771

Bray J (2008) Consumer behaviour theory: approaches and models

Brosdahl DJC, Carpenter JM (2010) Consumer knowledge of the environmental impacts of textile and apparel production, concern for the environment, and environmentally friendly consumption behavior. J Text Apparel Technol Manage 6:1–9

Browne MW, Cudeck R (1992) Alternative ways of assessing model fit. Sociol Methods Res 21:230–258

Butler SM, Francis S (1997) The effects of environmental attitudes on apparel purchasing behavior. Clothing Text Res J 15:76–85

Buttel FM, Flinn WL (1978) The politics of environmental concern: the impacts of party identification and political ideology on environmental attitudes. Environ Behav 10:17–36

Carrigan M, Attalla A (2001) The myth of the ethical consumer—do ethics matter in purchase behaviour? J Consum Mark 18:560–577

Chan K (2000) Market segmentation of green consumers in Hong Kong. J Int Consum Mark 12:7–24

Chan T-Y, Wong CWY (2012) The consumption side of sustainable fashion supply chain. Understanding fashion consumer eco-fashion consumption decision. J Fashion Mark Manage 16:193–215

Cottrell SP, Graefe AR (1997) Testing a conceptual framework of responsible environmental behavior. J Environ Educ 29:17

Davis R, Lang B (2012) Modeling the effect of self-efficacy on game usage and purchase behavior. J Retail Consum Serv 19:67–77

de Klerk HM, Lubbe S (2008) Female consumers' evaluation of apparel quality: exploring the importance of aesthetics. J Fashion Mark Manage 12:36–50

Diamantopoulos A, Schlegelmilch BB, Sinkovics RR, Bohlen GM (2003) Can socio-demographics still play a role in profiling green consumers? A review of the evidence and an empirical investigation. J Bus Res 56:465

Dickson MA (2001) Utility of no-sweat labels for apparel consumers: profiling label users and predicting their purchases. J Consum Aff 35:96–119

Du Preez R (2003) Apparel shopping behaviour—Part 1: Towards the development of a conceptual theoretical model. SA J Ind Psychol 29

Du Preez R, Visser EM (2003) Apparel shopping behaviour—Part 2: Conceptual theoretical model, market segments, profiles and implications. SA J Ind Psychol 29

Dunlap RE, van Liere KD (1981) Environmental concern: does it make a difference how it's measured? Environ Behav 13:651

Eckman M, Kadolph SJ, Damhorst ML (1990) Toward a model of the in-store purchase decision process: consumer use of criteria for evaluating women's apparel. Clothing Text Res J 8:13–22

Engel JF, Blackwell RD, Miniard PW (1986) Consumer behaviour. The Dryden Press, New York

Fornell C, Larcker DF (1981) Evaluating structural equation models with unobservable variables and measurement error. J Mark Res 18:39–50

Foxall G (2004). Psychology in Behavioural Perspective, Beard Books

Gimenez C, Sierra V (2013) Sustainable supply chains: governance mechanisms to greening suppliers. J Bus Ethics 116:189–203

Ha-Brookshire JE, Norum PS (2011) Willingness to pay for socially responsible products: case of cotton apparel. J Consum Mark 28:344–353

Hair JF, Ringle CM, Sarstedt M (2011) PLS-SEM: indeed a silver bullet. J Mark Theory Pract 19:139–152

Halepete J, Littrell M, Park J (2009a) Personalization of fair trade apparel: consumer attitudes and intentions. Clothing Text Res J 27:143–160

Halepete J, Littrell M, Park J (2009b) Personalization of fair trade apparel: consumer attitudes and intentions. Clothing Text Res J 27:143–160

Han T-I, Chung J-E (2014) Korean consumers' motivations and perceived risks toward the purchase of organic cotton apparel. Clothing Text Res J 32:235–250

Harry J, Hendee J, Gale R (1969) Conservation: an upper-middle class social movement. J Leisure Res 2:246–254

Hartlieb S, Jones B (2009) Humanising business through ethical labelling: progress and paradoxes in the UK. J Bus Ethics 88:583–600

Hassan L, Shaw D, Shiu E, Walsh G, Parry S (2013) Uncertainty in ethical consumer choice: a conceptual model. J Consum Behav 12:182–193

Howard JA, Sheth JN (1969) The theory of bayer behavior. Wiley, New York

Honnold JA (1981). Predictors of public environmental concern in the 1970s. In: Mann MD (ed.) Environmental Policy Formation–the impacts of values, ideology and standards 68-Lexington. Lexington Books

Hustvedt G, Bernard JC (2010) Effects of social responsibility labelling and brand on willingness to pay for apparel. Int J Consum Stud 34:619–626

Hyllegard KH, Yan R-N, Ogle JP, Lee K-H (2012) Socially responsible labeling: the impact of hang tags on consumers' attitudes and patronage intentions toward an apparel brand. Clothing Text Res J 30:51–66

Jain SK, Kaur G (2006) Role of socio-demographics in segmenting and profiling green consumers: an exploratory study of consumers in India. J Int Consum Mark 18:107–146

Jolibert AJP, Baumgartner G (1981) Toward a definition of the consumerist segment in France. J Consum Res 8:114–117

Jung S, Jin B (2014) A theoretical investigation of slow fashion: sustainable future of the apparel industry [electronic resource]. Int J Consum Stud 38:510–519

Kang J, Kim S-H (2013) What are consumers afraid of? understanding perceived risk toward the consumption of environmentally sustainable apparel. Fam Consum Sci Res J 41:267–283

Kang J, Liu C, Kim S-H (2013) Environmentally sustainable textile and apparel consumption: the role of consumer knowledge, perceived consumer effectiveness and perceived personal relevance. Int J Consum Stud 37:442–452

Kassarjian HH (1971) incorporating ecology into marketing strategy: the case of air pollution. J Mark 35:61–65

Kiezel E (2010) Konsument i jego zachowania na rynku europejskim. Warszawa, PWE

Kim H-S, Damhorst ML (1998) Environmental concern and apparel consumption. Clothing Text Res J 16:126–133

Kinnear TC, Taylor JR, Ahmed SA (1974) Ecologically concerned consumers: who are they? J Mark 38:20–24

Koszewska M (2010) CSR standards as a significant factor differentiating textile and clothing goods. Fibres Text Eastern Europe 18:14–19

Koszewska M (2011a) Social and eco-labelling of textile and clothing goods as means of communication and product differentiation. Fibres Text Eastern Europe 19:20–26

Koszewska M (2011b) The ecological and ethical consumption development prospects in Poland compared with the Western European countries. Comp Econ Res 14:101–123

Koszewska M (2012a) Global Global Consumer trends and textile and clothing innovations—implications for polish enterprises. The Transformation of Consumption and Consumer Behaviour. Institute for Market, Consumption and Business Cycles Research, Warsaw

Koszewska M (2012b) Impact of ecological and social sensitivity on the way of consumers' behavior in the textiles and clothing market. Handel Wewnętrzny 3

Koszewska M (2012c) Role of consumers' input into the development of innovations. Innovative trends in the textile and clothing industry and the needs of polish consumers. Fibres Text Eastern Europe 20:9–15

Koszewska M (2013) A typology of Polish consumers and their behaviours in the market for sustainable textiles and clothing. Int J Consum Stud 37:507–521

Koszewska M (2015) Life cycle assessment and the environmental and social labels in the textile and clothing industry. In: Muthu SS (eds.) Handbook of Life Cycle Assessment (LCA) of textiles and clothing, 1st edn. Woodhead Publishing

Koszewska M, Treichel A (2015) Ecological performance of selected fabrics as a factor in the subjective evaluation of their comfort. In: 15th AUTEX World Textile Conference, Bucharest, Romania

Krajnc D, Glavič P (2003) Indicators of sustainable production. Clean Technol Environ Policy 5:279–288

Lavorata L (2014) Influence of retailers' commitment to sustainable development on store image, consumer loyalty and consumer boycotts: proposal for a model using the theory of planned behavior. J Retail Consum Serv 21:1021–1027

LEA Wickett J, Gaskill LR, Damhorst ML (1999) Apparel retail product development: model testing and expansion. Clothing Text Res J 17:21–35

Lee N, Choi YJ, Youn C, Lee Y (2012) Does green fashion retailing make consumers more eco-friendly?: the influence of green fashion products and campaigns on green consciousness and behavior. Clothing Text Res J 30:67–82

Leonard-Barton D (1981) Voluntary simplicity lifestyles and energy conservation. J Consum Res 8:243–252

Liu X, Wang C, Shishime T, Fujitsuka T (2012) Sustainable consumption: green purchasing behaviours of urban residents in China. Sustain Dev 20:293–308

Maloney MP, Ward MP (1973) Ecology: let's hear from the people: an objective scale for the measurement of ecological attitudes and knowledge. Am Psychol 28:583–586

Marcoulides GA, Schumacker RE (1996) Advanced structural equation modeling: issues and techniques. Erlbaum, Mahwah

McDonald S, Oates C, Thyne M, Alevizou P, McMorland L-A (2009) Comparing sustainable consumption patterns across product sectors. Int J Consum Stud 33:137–145

McDonald S, Oates CJ (2006) Sustainability: consumer perceptions and marketing strategies Bus Strategy Environ (Wiley) 15, 157–170

McEachern MG, Carrigan M (2012) Revisiting contemporary issues in green/ethical marketing: an introduction to the special issue. J Mark Manage 28:189–194

Mohai P, Twight BW (1987) Age and environmentalism: an elaboration of the buttel model using national survey evidence. Soc Sci Q (University of Texas Press) 68:798–815

Mohr LA, Webb DJ, Harris KE (2001) Do consumers expect companies to be socially responsible? The impact of corporate social responsibility on buying behavior. J Consum Aff 35:45

Moisander J, Markkula A, Eräranta K (2010) Construction of consumer choice in the market: challenges for environmental policy. Int J Consum Stud 34:73–79

Moon KK-L, Lai CS-Y, Lam EY-N, Chang JMT (2014) Popularization of sustainable fashion: barriers and solutions. J Text Inst 106:939–952

Muthu SS, Li Y, Hu J-Y, Mok P-Y, Lin M (2013) Modelling and quantification of Eco-functional Index: the concept and applications of eco-functional assessment. Ecol Ind 26:33–43

Nicosia FM (1966) Consumer decision process. Prentice Hall

Neuman K (1986) Personal values and commitment to energy conservation. Environ Behav 18:53

Nicosia FM (1968) Advertising management, consumer behavior, and simulation. J Advertising Res 8:29–37

Niinimäki K (2010) Eco-clothing, consumer identity and ideology. Sustain Dev 18:150–162

Ogrean C, Herciu M (2014) arguments for CSR-based sustainable competitiveness of multinationals in emerging markets (Part I). Stud Bus Econ 9:57–67

Olsson U (1979) Maximum likelihood estimation of the polychoric correlation coefficient. Psychometrika 44:443–460

Osterhus TL (1997) Pro-social consumer influence strategies: when and how do they work? J Mark 61:16

Ostman RE, Parker JL (1987) Impact of education, age, newspapers, and television on environmental knowledge, concerns, and behaviors. J Environ Educ 19:3

Ozcaglar-Toulouse N, Shaw D, Shiu E (2006) In search of fair trade: ethical consumer decision making in france [electronic resource]. Int J Consum Stud, 30:502–514

Paço A, Alves H, Shiel C, Filho WL (2013) Development of a green consumer behaviour model. Int J Consum Stud 37:414–421

Pickett GM, Kangun N, Grove SJ (1993) is there a general conserving consumer? A public policy concern. J Public Policy Mark 12:234–243

Pino G, Peluso AM, Guido G (2012) Determinants of regular and occasional consumers' intentions to buy organic food. J Consum Aff 46:157–169

Plesniak A (2006) Identyfikacja zależności korelacyjnej w procedurze SEM. Ilościowe i jakościowe metody badania rynku. Pomiar i jego skuteczność. Poznan, Zeszyty Naukowe AE w Poznaniu

Pookulangara S, Shephard A (2013) Slow fashion movement: understanding consumer perceptions—an exploratory study. J Retail Consum Serv 20:200–206

Ray JJ (1975) Measuring environmentalist attitudes. Aust New Zealand J Sociol 11:70–71

Roberts JA (1996) Green Consumers in the 1990s: profile and Implications for advertising. J Bus Res 36:217–231

Romli A, Prickett P, Setchi R, Soe S (2015) Integrated eco-design decision-making for sustainable product development. Int J Prod Res 53:549–571

Sagan A (2011) Modele strukturalne w analizie zachowań konsumenta– ewolucja podejść. Konsumpcja i Rozwój 1

Samdahl DM, Robertson R (1989) Social determinants of environmental concern: specification and testof the model. Environ Behav 21:57

Schlegelmilch BB, Bohlen GM, Diamantopoulos A (1996) The link between green purchasing decisions and measures of environmental consciousness. Eur J Mark 30:35

Scott D, Willits FK (1994) Environmental attitudes and behavior: a Pennsylvania survey. Environ Behav 26:239–260

Shaw D, Shiu E, Hassan L, Bekin C, Hogg G (2007) Intending to be ethical: an examination of consumer choice in sweatshop avoidance. Adv Consum Res 34:31–38

Shim S, Kotsiopulos A (1992) Patronage behavior of apparel shopping: part II. testing a patronage model of consumer behavior. Clothing Text Res J 10:58–64

Shrum LJ, McCarty JA, Lowrey TM (1995) Buyer characteristics of the green consumer and their implications for advertising strategy. J Advertising 24:71–82

Smyczek S, Sowa I (2005) Konsument na rynku. Zachowania, modele, aplikacje, Warszawa, Difin

Staniškis J, Arbačiauskas V, Varžinskas V (2012) Sustainable consumption and production as a system: experience in Lithuania. Clean Technol Environ Policy 14:1095–1105

Tai SHC, Tam JLM (1996) A comparative study of Chinese consumers in asian markets-a lefestyle analysis. J Int Consum Mark 9:25–42

Thilak V, Saravanan D (2015) Eco-design/sustainable Design of textile products. Handbook of sustainable apparel production. CRC Press

Thogersen J (2000) Psychological determinants of paying attention to eco-labels in purchase decisions: model development and multinational validationnn. J Consum Policy 23:285–313

Thøgersen J (2000) Psychological determinants of paying attention to eco-labels in purchase decisions: model development and multinational validation. J Consum Policy 23:285–313

Tognacci LN, Weigel RH, Wideen MF, Vernon DTA (1972) Environmental quality: how universal is public concern? Environ Behav 4:73–86

Trudel R (2011) Socially conscious consumerism. Network for Business Sustainability. [WWW document]. Accessed on 10th July, 2013 URL http://www.nbs.net/wp-content/uploads/NBS-Consumerism-Primer1.pdf

Tucker JLR, Dolich IJ (1981) Profiling environmentally responsible consumer-citizens. J Acad Mark Sci 9:454

Vignali C (1999) Benetton's brand position explored and developed through Nicosia's consumer-behaviour model. J Text Inst 90:48–59

Visser EM, Du preez R (2001) Apparel shopping orientation: two decades of research. J Fam Ecol Consum Sci 2

Webster FE Jr (1975) Determining the characteristics of the socially conscious consumer. J Consum Res 2 188–196

Young W, Hwang K, McDonald S, Oates CJ (2010) Sustainable consumption: green consumer behaviour when purchasing products. Sustain Dev 18:20–31

Zimmer MR, Stafford TF, Stafford MR (1994) Green issues: dimensions of environmental concern. J Bus Res 30:63–74

The Feasibility of Large-Scale Composting of Waste Wool

Gwendolyn Hustvedt, Erica Meier and Tina Waliczek

Abstract Although wool remains a popular fiber due to its durability, comfort, and ease of quality production, increasing the sustainability of wool fashion products requires attention to all portions of the wool life cycle, including production of sheep to provide the wool. Managing sheep in many regions of the globe results in excess or waste wool that is not suitable for the textile or fashion supply chain. One way to increase the economic, environmental, and social sustainability of wool production is to compost the waste wool into soil amendments and landscaping aids that can provide producers with additional income, provide horticulturalists with an environmentally friendly alternative to other soil amendments, and provide communities that depend on sheep a pride in the value of all parts of the life cycle. This chapter outlines an experimental study that was conducted to determine the proper proportions of sheep waste products to other biomass that would be needed in a large-scale composting operation in order to produce a high-quality compost valuable to the horticulture and agriculture industries. The results of an experimental trial with waste wool determined that a 25 % waste wool, 50 % grass clipping, and 25 % horse stall waste mixture provided the optimal results for composting in a large-scale manner. Separation of compacted wool, if transported in wrapped bundles, proved essential for allowing sufficient decomposition of the waste wool. The composting produced was tested and determined to be of acceptable quality.

G. Hustvedt (✉)
School of Family and Consumer Sciences, Texas State University,
601 University Drive, San Marcos, TX 78666, USA
e-mail: gh21@txstate.edu

E. Meier · T. Waliczek
Department of Agriculture, Texas State University,
601 University Drive, San Marcos, TX 78666, USA

T. Waliczek
e-mail: tc10@txstate.edu

© Springer Science+Business Media Singapore 2016
S.S. Muthu and M.A. Gardetti (eds.), *Green Fashion*,
Environmental Footprints and Eco-design of Products and Processes,
DOI 10.1007/978-981-10-0111-6_4

Keywords Waste   management · Life   cycle · Biomass · Wool · Sheep · Composting

1 Introduction

As a natural fiber (i.e., a fiber that occurs naturally in the form of a usable fiber for textile production), wool has played a role in clothing for human beings since before fashion or the fashion industry existed. Ninety percent (90 %) of all wool production occurs in under 20 countries, however, even within these countries, the millions of small holdings and family operations vastly outnumber the 200,000 commercial operations and the operations in the countries that produce the remaining 10 % are not large enough to warrant representation in the international wool organizations (i.e., Ireland or the Faroe Islands; IWTO 2015). Although sheep and wool production has become industrialized in many regions of the world, just as in all other parts of agricultural production, the raising of sheep and the harvesting of wool remains an important part of the social, economic, and environmental systems in many cultures.

In the 1940s, the United States was the world's fifth largest wool-producing nation (National Research Council 2008), and the state of Texas is the United States' leading wool producer, with the main sheep production areas of Texas in the Edwards Plateau in the west of Texas. Wool production in Texas is part of a cyclical market system that includes the marketing of fiber and meat. When consumer demand drives the price of meat high enough to surpass the value of the fiber, producers will slaughter their fiber stock for meat, reducing the supply of fiber (Jones 2004). The price of wool, however, depends mainly on orders from apparel merchandisers and in the past decade fashion is cycling more quickly than the sheep industry can respond to the changes in demand. Finding alterative outlets for fiber that are more dependable than consumer interest in wool apparel could provide base support for Texas wool prices.

Composting is the natural breaking down of carbon- and nitrogen-containing materials such as recycled plant parts, food scraps, paper, animal fodder, and wood chips. The end product, compost, becomes a beneficial soil amendment because it contains a full spectrum of essential plant nutrients that improve soil productivity (Emerson 2003). Wool has historically been used in garden beds and anecdotal records from gardeners from the 1940s suggest beneficial effects for plants. Wool is a natural source of nitrogen and potassium and acidifies the soil, improving the pH for the growth of many garden and greenhouse crops (Poston 2006). In central Texas, a garden soil amendment that would hold moisture and lower the pH of the highly alkaline native soils has a potentially high market value.

This chapter outlines an experimental study that was conducted to determine the proper proportions of sheep waste products to other biomass that would be needed in a large-scale composting operation in order to produce a high-quality compost valuable to the horticulture and agriculture industries. The first portion of this chapter is a literature review that discusses the source of waste from wool production, with a specific focus on waste fibers. Next, the literature review examines previous research on composting in general and the composting of waste wool in particular. The next portion of the chapter lays out how the waste wool was processed into several types of compost, with the goal of determining the best methods for dependably producing quality compost. Finally, the results of the monitoring during the compost process as well as the results of quality analysis on the finished product are laid out and discussed. This topic is important because the sustainability of wool production, especially from sheep production over large regions such as West Texas, requires attention to each stage of the life cycle. The disposal of wool that is not suitable and/or not required by the textile industry is one aspect of this life cycle and the value of this study is that it describes the formulation of the composting technique in a way that it can be used by those seeking to service larger fiber production systems with large-scale composting operations.

2 Literature Review

Fiber-producing animals such as sheep, goats, and camelids (llamas and alpacas) are browsers that can contribute to the health of an organically or sustainably managed pasture. By eating the plants that do not appeal to cattle, browsers make pastures easier to manage (Animut and Goetsch 2008). Parasite control for pastured livestock can also be improved if different species are alternated through pastures, presenting less desirable hosts for species-specific worms (Hermansen 2003). The incorporation of fiber-producing animals would enhance the overall system of sustainable livestock production as well by providing an additional source of income for small producers if they are able to produce fiber. Sometimes, however, the fiber produced by sheep is not of high enough quality to be suitable for use in woven or knitted textiles.

Breeds of sheep called hair sheep are becoming popular in Texas because hair sheep are covered with hair instead of wool, diverting nutrients from the production of a dense woolly fleece to better meat production (Wildeus 1997). Ranchers have shifted their production of sheep towards hair sheep in part because of unstable wool prices compared to the relative stability of meat prices. Because wool-producing sheep are often run in pasture alongside hair sheep, some Texas sheep producers raise sheep that are crossbred wool and hair sheep that grow random patches of wool. Wool shorn from these sheep for their comfort and health is mixed with coarse hair and is not considered high quality but rather considered "waste," and is, therefore, readily available and relatively inexpensive. In other parts of the world, where sheep are raised primarily for meat (e.g., Faroe Islands), the quality of the wool is not necessarily of primary importance and selective breeding, where it occurs, would focus on the live weight at slaughter rather than fleece quality. This means that any solution that provided sustainable disposal of waste wool, such as large-scale composting, would be suitable not just for West Texas, but also for other regions.

In addition to the production of "waste" wool from crossbred or "meat" sheep, the fleece cut from certain areas of the wool-producing sheep's body, including the hind and underside areas, is typically stained and matted with waste from defecation or from soil and vegetation. This wool, called "tags," is also often contaminated with weed seeds. Like waste wool, tags are also considered of poorer quality and sold at lower prices than wool shorn from other areas of the sheep's body. One benefit of composting waste wool or tags would be the removal of weed seeds that currently could prevent direct use of the wool as mulch. As a waste-management system within agriculture, composting is known to kill weed seeds if temperatures are high enough and maintained for long periods of time. For germination to be inhibited on weed seeds of sorghum, bindweed, pigweed, johnsongrass, and kochia, temperatures of 120–180 °F for 3–7 days on average must be obtained to kill the seeds (Wiese et al. 1998). Another source of waste in the sheep industry comes from wool scouring, a cleaning process (Phillips 1936). Raw wool is approximately 50 % grease by weight. Removing this grease, called lanolin, produces large quantities of sludge, which then poses a waste disposal problem for the scouring facilities. This sludge is not toxic (unless the soil used to graze sheep contains compounds such as arsenic), however, it becomes rancid quickly and must be properly disposed of, adding to the cost of wool production.

Recent research found that wool-incorporated plant potting improves the water-holding capacity of the soil and acts as a slow-release fertilizer (Górecki and Górecki 2010). In a study conducted in the United Kingdom, research found that utilizing wool scouring sludge waste within composting "produced a safe, useful, and marketable product" (Pearson et al. 2004, p. 9). Zheljazkov (2005) found a surplus of wool in storage in Atlantic Canada from previous harvests due to the lack of demand for wool, and therefore, no longer a commercially viable product. As a result, an alternative use for this protein-rich product and its by-products was needed. Zheljazkov (2005) demonstrated that wool and wool wastes decomposed slowly under controlled conditions and acted as a slow-release fertilizer, releasing N, P, K, and S into the soil. Wool is also hygroscopic, meaning that it readily takes up and retains moisture (Mirriam-Webster 1995), where wool generally retains a 15–20 % moisture content.

The literature also suggests that wool-based composts acidify soils, which is a desirable effect for soils that are predominantly alkaline (Zheljazkov et al. 2009). The drought-affected dry alkaline soils of central Texas would be appropriate for observing and examining the nutrient-rich, water-holding, acidifying wool-based compost as a soil amendment. Furthermore, the quality of composts depends primarily on the basic raw materials composted, meaning if the basic raw materials are of high quality then ultimately the resulting compost is poised to be of high quality. Compost in general has already been identified as having beneficial effects on soil properties, including increased soil aggregation, soil aeration and permeability, erosion control, water-holding capacity, and an effective biological weed control agent, as well as decreased plant pathogens and diseases.

The incorporation of wool as a compost feedstock has the potential to create compost of high quality while effectively utilizing unwanted wool, as well as producing a product valuable to the horticultural and agricultural industries. Previous studies such as Zoccola et al. (2014), although explaining the importance of composting to handling the issue of waste wool, do not provide enough detail on the appropriate techniques and formulation to facilitate adoption of large-scale composting as a method. The goal of this chapter, therefore, is to provide clear comprehensive information on a method that can be used for the large-scale composting of waste wool into a value-added soil amendment.

3 Method

3.1 Compost Site

Compost piles were constructed at the Bobcat Blend composting site at the Mueller Farm owned by Texas State University. Texas State University Muller Farm was previously utilized as an alternative grazing source for the livestock kept at Texas State University Freeman Ranch and is approximately 125 acres. Of the 5 acres allocated for the compost site, 2.285 acres were transformed into a catchment pond that could withstand a 25-year 24-h flood event. The remaining 2.715 acres were cleared and graded so that any water run-off from the compost piles would be captured by the catchment pond. Fences and gates were also installed to keep out any livestock and to contain the feedstocks utilized for composting.

3.2 Wool Waste

During the fall of 2010, four bundles of wool waste and tags were obtained from sheep producers in the San Angelo, Texas area. The wool waste and tags came in highly compacted bundles that were roughly one meter wide, one and a half meters in length, and one meter in height, and each bundle weighed approximately 227 kg. Wool waste is any wool not deemed marketable by those conducting shearing and is of variable quality. Tags are parts of the wool fleece cut from certain areas of the sheep's body that include the underside and hind areas, which are usually stained from vegetation, soil, or defecation as well as contaminated with weed seeds and are considered of poorer quality. However, a benefit of composting wool tags and waste would be the elimination of any weed seeds due to the high temperatures of the composting process. This lower quality wool is considered a waste product by the wool industry but a potential nitrogen feedstock within a compost operation, producing a wooly compost that is a high-quality compost valuable to both the agriculture and horticulture industries.

Table 1 Compost pile composition by weight and source

Pile		Wool	Carbon	Other Nitrogen
	% by Volume	25 %	50 %	25 %
		kilos	kilos	kilos
A		227	1742—woodchips	1087—horse stall waste (manure/sawdust)
B		227	1742—woodchips	653—water hyacinth (*Eichhornia crassipes*)
C		227	1742—woodchips	1361—food waste
D		227	1687—dried grass clippings	1087—horse stall waste (manure/sawdust)

3.3 Compost Mixtures

In order to create compost piles of ideal conditions for composting and to achieve high enough temperatures to kill weed seeds as well as pathogens, various percentages of other compost feedstocks were utilized and altered as needed. Wiese and colleagues (1998) reported that the composting process has been known to reach high temperatures for long periods of time, where "several days of pile temperatures above 54.4 °C (130.0 °F) … [destroys] pathogens and weed seeds" (Dougherty 1999, p. 47). Previous studies on the use of wool waste in the composting process suggested a recipe of 39 % wool waste (nitrogen feedstock) and 61 % woodchips (carbon feedstock) (Das et al. 1997). This recipe was manipulated for this study to create an essentially 50/50 mixture of carbon and nitrogen sources with the wool being balanced by other nitrogen feedstock sources to build piles that were 25 % wool, 50 % woodchips, and 25 % various other nitrogen feedstock sources (see Table 1).

The first pile (Pile A) contained by volume 25 % wool waste, 25 % horse stall waste (approximately 80 % manure and 10 % sawdust), and 50 % woodchips. The mixture was monitored and turned every 5–7 days. Previous experience at the facility suggested that composting should have been completed within 30 days (Rynk 1992). However, the wool waste and tags were highly compacted within all four bundles and when this first bundle was incorporated into Pile A, separation of these two types of material was minimal. For this reason, larger clumps of wool waste were present after 15 days and it became apparent that there was minimal decomposition of these clumps. Consequently, the composting process took longer than the customary 30 days, and the pile continued to be turned every 5–7 days until the desired compost texture was obtained.

Consultation of the literature confirmed that wool waste in the fibrous form had indeed previously been composted (Pearson et al. 2004). However, for the three remaining compost piles that were constructed, separation of the wool waste and tags seemed to be desirable in order to obtain a more fibrous, less clumped nature for the wool material. Separation of the compacted wool waste and tags of each bundle was done by hand, where the bundle was opened up and pulled apart manually. The manual separation process took several hours for each bundle and was thus split up into 3–4-h sessions over the span of a couple of days. Ultimately, the separation of each bundle took approximately 4 days and the separated wool waste was stored in twelve 55-gallon bins until the entire bundle was processed and ready for compost pile construction.

The hygroscopic nature of wool suggested that it would be important to consider the moisture profile of the feedstock in order to allow for enough moisture to be available for the decomposition process compared with the moisture likely to be absorbed and held within the fibers. Additional consultation of the literature suggested that the inclusion of Bracken fern (*Pteridium aquilinum*) within composting formulas that included sheared wool waste would create a nutrient-rich soil amendment (Poston 2006). Bracken ferns are a wetlands indicator plant that require well-drained but high levels of moisture to grow, and consequently have a relatively significant weight by water content (Rook 2008; United States Department of Agriculture [USDA] n.d.). Likewise, research by Montoya and colleagues (2013) documented the success of composting the invasive aquatic species water hyacinth (*Eichhornia crassipes*) which, due to its weight by water content, had the potential to substitute for the Bracken fern (*Pteridium aquilinum*) as a compost feedstock (Martin and Gershuny (eds.) 1992). By utilizing a high water content feedstock, there was the anticipation for the need to add less water to the pile in order to maintain the ideal moisture content. For that reason, the second compost pile (Pile B) was constructed utilizing the invasive aquatic species water hyacinth (*Eichhornia crassipes*), and the recipe consisted of 25 % wool waste, 25 % water hyacinth (*Eichhornia crassipes*), and 50 % woodchips.

The Texas State University Bobcat Blend program has been collecting and implementing cafeteria food waste as a nitrogen feedstock within other composting research projects. Although food waste is less likely to be available in the less densely populated regions where sheep are typically run, it was decided to include this type of high-quality nitrogen source. The third compost pile (Pile C) was constructed utilizing the cafeteria food waste and the recipe consisted of 25 % wool waste, 25 % food waste, and 50 % woodchips.

Due to the slower decomposition rate of the wool material, the fourth compost pile (Pile D) was constructed with consideration to texture as compared with the other compost feedstocks. The larger particulate feedstocks such as woodchips were observed to take longer to mix in with the wool material, and needed turning of the pile over a 2–3-week period before materials were blended together enough for decomposition to begin. The smaller particulate feedstocks were observed to mix and blend in with the wool material easier, and likely initiated decomposition at an earlier stage compared to the larger particulates. In addition, previous research has found that smaller particulate organic substances were more easily degradable and decomposition was more rapid, suggesting a higher degree of microorganism activity in the smaller particulate mixtures (Hoppenheidt et al. 2000). Therefore, the fourth compost pile (Pile D) was constructed utilizing a recipe that consisted of 25 % wool waste, 25 % horse stall waste (approximately 80 % manure and 10 % saw dust), and 50 % dried grass clippings. The smaller particulate dried grass clippings, an acceptable carbon feedstock, were utilized as a substitute for the bigger particulate woodchip feedstock, and the horse stall waste was retained to allow comparison with Pile A which provides both a baseline and easily reproduced composition.

Compost piles were constructed based on procedures described by Rynk and colleagues (1992), and were approximately 1.5–1.8 m (5–6 ft) tall and 3.0–3.5 m (10–12 ft) wide. This height and width allowed the piles to be insulated and generate enough heat to kill pathogens (including weed seeds and pathogens). Turning was conducted weekly using a "bobcat" front-end loader to ensure that formerly outer exposed surfaces were buried within the pile each time the pile was turned. This allowed all weed seeds and pathogens the opportunity to be exposed to high temperatures of the interior of the piles.

3.4 Compost Pile Monitoring

Each pile was maintained, monitored, and recorded consistent with the composting procedures practiced thus far and are described below. The temperature, pH, moisture, and maturity of each pile was monitored, maintained, and recorded by the researcher. A moisture level of 50–60 % is ideal (Rynk 1992). Moisture levels were measured with a 60-in. Compost Moisture Meter (ReoTemp Instrument Corporations, San Diego, California), as well as with a "feel" test. A feel test involves taking a handful of compost and squeezing it to observe whether it feels like a moist sponge; if the sample does not feel wet to the touch, then it is too dry and if water can be squeezed out then it is too wet (Rynk 1992). However, the wool waste required saturation in order to absorb water due to the wool's initial tendency to repel water and then to absorb water, which required a significant amount of extra watering in order to maintain a moisture level of 50–60 %. Acidity and alkalinity (pH) were measured and recorded with a hand-held Kelway® Soil pH sensor (Wyckoff, New Jersey). The temperatures of the compost piles were monitored and measured with the 60-in. Fast Response Compost Thermometer. Oxygen levels were monitored and measured with an oxygen monitor (Model No. 0-21, Demista Instruments, Arlington Heights, Illinois). Temperature, oxygen, pH, moisture, and maturity levels of each pile was also monitored, recorded, and maintained through the scheduled turning of the piles, with the achieved goal that the piles remained aerobic for best microbial activity.

4 Results

4.1 Compost Pile Progression

Construction of Pile A utilized approximately 227 kg of wool waste material, 1087 kg of the horse manure mixture, and 1742 kg of woodchips. The average pH

Table 2 Initial field results of compost pile measurements

Pile	Composition	Moisture (%)	Temperature (°C)	pH
A	Wool, manure and woodchips	89.3	65	7.00
B	Wool, water hyacinth and woodchips	65.0	54	7.00
C	Wool, food waste and woodchips	53.0	63	7.00
D	Wool, manure and grass clippings	71.0	67	6.90

of the Pile A was recorded at 7.01 which falls within the acceptable range of 5.0–8.5 (Rynk 1992; see Table 2). The average moisture content of Pile A was 89.3 % which is relatively high considering a moisture level of 50–60 % is ideal (Rynk 1992), and could potentially explain why the composting period of Pile A took longer. However, even though the Pile A required a longer composting period, it achieved a high average temperature of of 65.1 °C (149.1 °F), and, after 3 months, the pile still maintained an average temperature of 61.9 °C (143.5 °F).

Construction of Pile D utilized approximately 227 kg of wool waste material, 653 kg of water hyacinth (*Eichhornia crassipes*) plants, and 1742 kg of woodchips. The average pH of Pile B was recorded at 7.03 which also falls within the acceptable range of 5.0–8.5 (Rynk 1992). The average moisture content of Pile B was 65.0 % which is only slightly higher when compared to the ideal moisture content of 50–60 % (Rynk 1992). On the other hand, Pile B hardly reached high enough temperatures for the composting process to be successful in killing weeds and pathogens. It only achieved a high average temperature of 53.8 °C (128.8 °F). However, the temperature was also constant, because after 3 months the pile still maintained an average temperature of 52.4 °C (126.2 °F).

Construction of Pile C utilized approximately 227 kg of wool waste material, 1362 kg of food waste, and 1742 kg of woodchips. The average pH of the Pile C was similarly recorded at 7.02 which also falls within the acceptable range of 5.0–8.5 (Rynk 1992). The average moisture content of Pile C was 52.6 % which fell within the ideal moisture content of 50–60 % (Rynk 1992). Pile C achieved a high average temperature of 63.1 °C (145.5 °F), where even after 3 months it still maintained an average temperature of 55.8 °C (132.5 °F).

Construction of Pile D utilized approximately 227 kg of wool waste material, 1087 kg of horse manure mixture, and 1687 kg of grass clippings. The average pH of the Pile D was slightly lower, recorded at 6.9 which still falls within the acceptable range of 5.0–8.5 (Rynk 1992). The average moisture content of Pile D was also slightly higher than the ideal moisture level of 50–60 %, maintaining an average moisture content of 71.1 %. However, the composting period of Pile D was significantly more rapid (potentially due to the smaller particulate mixture) where it achieved a high average temperature of 66.8 °C (152.3 °F), where even after 3 months the pile had still maintained an average temperature of 60.8 °C (141.4 °F).

The compost piles were observed and examined by sight and touch, as well as with moisture meter and thermometer readings to ensure that the compost was decomposed and suitable for the curing stage. The curing stage occurs when there is a sustained reduction in pile temperatures (10.0–40.6 °C/50.0–105.0 °F) (Rynk 1992). The compost was cured in the same piles where they were built. The curing stage generally takes approximately 4–8 weeks (Rynk 1992). Ultimately, although the decomposition of the wool waste material took longer than expected, the curing period was consistent with the conclusion made by Rynk (1992).

Because no additional feedstock was required to maintain decomposition, a total of approximately 11,521 kg of waste materials were utilized as compost feedstocks; thus approximately 11,521 kg of waste material was diverted from landfills.

4.2 Compost Quality Test

After curing, the woolly compost was screened by hand, where hand screening as opposed to using a wire-mesh screener, allowed for the removal of larger particles and any foreign materials without removing the woolliness of the compost. Samples of the compost were then sent for analysis to the Agricultural Analytical Services Laboratory (Pennsylvania State University, University Park, Pennsylvania), where they were subjected to the laboratory's Compost Tests for U.S. Compost Council's certified Seal of Testing Approval Program. Because piles of compost often vary from location to location within the pile, the samples collected for analysis must be representative of all material being analyzed (Pennsylvania State University 2002). Therefore, the compost samples sent for analysis were composite samples of material collected from several different locations and depths within the pile being sampled. From each compost pile, samples were collected using recommendations from the Pennsylvania State University Agricultural Analytical Laboratory, where samples were collected at five locations and from three different depths from each location. Therefore, 15 samples were collected from each pile, where a total of 60 samples were collected, combined, and intermixed to create the composite samples sent for analysis. The Agricultural Analytical Services Laboratory (Pennsylvania State University, University Park, Pennsylvania) determine and designate quality standards for the industry (Pennsylvania State University 2002). There, the Compost Tests for U.S. Compost Council's certified Seal of Testing Approval Program analyzed the compost and test for pH, moisture content, soluble salt content, and nutrient content [bioassay (maturity), total organic matter, total carbon (C), total nitrogen (N), calcium (Ca), magnesium (Mg), phosphorus (P), potassium (K)], along with any trace metals [arsenic (As), cadmium (Cd), copper (Cu), lead (Pb), mercury (Hg), molybdenum (Mo), nickel (Ni), selenium (Se), and zinc (Zn)], as well as stability (respirometry), particle size, pathogens (fecal coliform, *Coliform bacillus* and/or salmonella, *Salmonella* spp.), and weed seed(s) viability (Pennsylvania State University 2002; United States Composting Council 2010; see Table 3).

Table 3 Laboratory tests of 60 combined samples (15 from each pile)

Analyte	Results (as is) range	Results (dry weight) range	Typical range
PH	7.2–7.5	–	5–8.5
Soluble salts	2.24–2.51 mmhos/cm	–	1–10 mmhos
Solids	**59.4–62.9 %**	–	50–60 %
Moisture	**37.1–40.6 %**	–	40–50 %
Organic matter	15.6–18.6 %	**25.2–31.3 %**	30–70 % (dry weight)
			50–60 % ideal
Total nitrogen	0.006	0.9–1.1 %	<1–5 % (dry weight)
Organic nitrogen	0.60 %	0.9–1.1 %	<1–5 % (dry weight)
Ammonium N	0.0041–0.0072 %	0.0068–0.114 %	<1–5 % (dry weight)
Carbon	7.9–8.9 %	12.7–15.0 %	Varies
CN ratio	13.80–14.10	13.80–14.10	Varies (<20 ideal)
Phosphorus	0.24–0.31 %	0.39–0.52 %	Varies
Potassium	0.35–0.38 %	0.59–0.60 %	Varies
Calcium	5.05–5.36 %	8.45–8.52 %	Varies
Magnesium	0.20–0.22 %	0.32–0.34 %	Varies
Particle size	85.94–92.54 %	–	–

4.3 Implications of Results

The results of this testing indicate that, on average, the compost produced from waste wool was suitable for use in horticulture and agriculture. The percentage of organic matter was at the low end of the acceptable range and 20 % below the ideal range. This suggests that wool waste, in and of itself, is likely not a sufficient source of nitrogen. Future exploration of compost pile formulation should decrease the percentage of wool waste in comparison to other nitrogen sources (e.g., other livestock manure).

The percentage of solids was on the high end of the typical range and the moisture was on the low end of the typical range. The high percentage of solids is likely due to the clumped nature of the waste wool bundles and may possible be remedied if the waste and tags are separated at the shearing house and if the bundles are not so highly compacted. The co-mingling and compaction may be desirable if the intended disposal is a landfill that accepts waste by weight and decreasing the volume allows for greater efficiency during transport. However, if the intended disposal is large-scale composting, a waste operator should be motivated to reward production of less compacted wool waste. Future research should examine methods of mechanically separating compacted waste wool bundles. In terms of the other results, the low moisture content is less than ideal, but is most likely due to the drought conditions that were prevalent during the study and/or the drying out of the sample during shipping. This issue would be avoided in a commercial operation if the compost were sold as soon as the desired level of decomposition was reached or compost was remoistened periodically by rains or irrigation.

Overall, the results of this study suggest that the composition of feedstocks used in Pile D would provide the best results for large-scale composting of waste wool in terms of ease of production. In regions where fresh grass clippings are not easily obtained, leaf mold would provide a suitable substitute in terms of the moisture content and particulate size. The separation of compacted waste wool into smaller, more fibrous bundles is desirable where non-compacted waste wool proves infeasible to transport to the composting site.

5 Conclusion

The management of low quality or fecal-contaminated wool may not strike some as a significant question for the fashion industry, but it targets the root of the value of life-cycle analysis. If sustainability includes the economic well-being of participants in the chain of value, should not the efforts to improve economic viability of multi-species ranching in regions where the use of browsers along with grazers creates improvements in the soil health and plant biodiversity? Ranchers are not in the position of suggesting that the health and care of animals should be subject to the whims of the fashion consumer; however, without consistent markets for animal fibers, sheep ranchers are put in the awkward position of disposing of wool not desired by the market in the hopes of being ready to provide wool when it is required by the fashion industry. The cultural aspect of sustainability also comes into play, given that in many parts of the world, sheep are not being produced industrially, but are part of biodiverse, family-run operations where respect for the animal can be increased if unmarketable wool is not simply burned but incorporated into the other aspects of the farm ecosystem. The small experiment outlined in this chapter was an attempt to understand the viability of using waste wool for nonfashion purposes in order to reduce the environmental, economic, and cultural weight of caring for sheep when they are not "in style".

References

Animut G, Goetsch AL (2008) Co-grazing of sheep and goats: benefits and constraints. Small Rumin Res 77(2):127–145

Das KC, Tollner EW, Annis PA (1997) Bioconversion of wool industry solid wastes to value-added products. http://hdl.handle.net/1853/10336. Accessed 31 Aug 2015

Dougherty M (ed) (1999) Field guide to on-farm composting. Natural Resource, Agriculture, and Engineering Service, New York

Emerson D (2003) Building strong markets for mulch and compost products. Biocycle 44(7):36–40

Górecki RS, Górecki MT (2010) Utilization of waste wool as substrate amendment in pot cultivation of tomato, sweet pepper, and eggplant. Pol J Environ Stud 19:1083–1087

Hermansen JE (2003) Organic livestock production systems and appropriate development in relation to public expectations. Livest Prod Sci 80(1):3–15

Hoppenheidt K, Kottmair A, Mücke W et al (2000) Influence of organofluorine-treated textiles on biowaste composting. Text Res J 70(1):84–90

International Wool Textile Organization (2015) Facts. http://www.iwto.org/about-iwto/vision-mission/. Accessed 31 Aug 2015

Jones KG (2004) Trends in the U.S. Sheep Industry. United States Department of Agriculture, Economic Research Service, Agricultural Information Bulletin Number 787

Martin DL, Gershuny G (eds) (1992) The Rodale book of composting. Rodale Press Inc., New York

Montoya J, Waliczek TM, Abbott M (2013) Large-scale composting as a means of managing water hyacinth, Eichhornia crassipes. Invasive Plant Sci Manag 6(2):243–249

National Research Council (2008) Changes in the Sheep Industry in the United States. http://dels.nas.edu/resources/static-assets/materials-based-on-reports/reports-in-brief/SheepFinal.pdf. Accessed 31 Aug 2015

Pearson J, Lu F, Gandhi K (2004) Disposal of wool scouring sludge by composting. Department of Textiles. University of Huddersfield, United Kingdom

Phillips H (1936) Some fundamental principles of raw wool washing. J Textile Inst Proc 27(7):208–218

Pennsylvania State University (2002) Compost analysis: Sampling and mailing procedure. Agricultural Analytical Services Laboratory. Pennsylvania State University University Park Pennsylvania

Poston T (2006) Using the wool no-one wants. http://news.bbc.co.uk/2/hi/5317358.stm. Accessed 31 Aug 2015

Rook E (2008) Pteridium aquilinum, bracken fern. Ferns and Fern Allies of the Boundary Waters Wilderness and North Woods. http://www.rook.org/earl/bwca/nature/ferns/pteridiumaqui.html. Accessed 31 Aug 2015

Rynk R (1992) On-farm composting handbook. National Resource, Agriculture, and Engineering Service, New York

United States Composting Council (2010) Seal of testing assurance. http://www.compostingcouncil.org. Accessed 26 Feb 2010

United States Department of Agriculture [USDA] (n.d.) Profile for Pteridium aquilinum (western brackenfern). National Plant Data Center. Natural Resources Conservation Service. United States Department of Agriculture. http://plants.usda.gov/. Accessed 26 Feb 2010

Weise A, Sween J, Bean B et al (1998) High temperature composting of feedlot manure kills weed seeds. Am Soc Agric Biol Eng 14(4):377–380

Wildeus S (1997) Hair sheep genetic resources and their contribution to diversified small ruminant production in the United States. J Anim Sci 75(3):630–640

Zheljazkov VD (2005) Assessment of wool waste and hair waste as soil amendment and nutrient source. J Environ Qual 34(6):2310–2317

Zheljazkov VD, Stratton Pincock GW, Butler J et al (2009) Wool-waste as organic nutrient source for container-grown plants. Waste Manag 29(7):2160–2164

Zoccola M, Montarsolo A, Mossotti R et al (2014) Green hydrolysis conversion of wool wastes into organic nitrogen fertilisers. In: 2nd international conference on sustainable solid waste management, pp 1–11

Do as You Would Be Done by:
The Importance of Animal Welfare
in the Global Beauty Care Industry

Nadine Hennigs, Evmorfia Karampournioti and Klaus-Peter Wiedmann

Abstract Nowadays, the concept of sustainability is discussed in almost every product category. In this context, companies commit themselves to advancing good social, environmental, and animal-welfare practices in their business operations, including sustainable sourcing practices. Nevertheless, even if many companies in the global beauty care industry have embraced such claims, common practices such as water pollution, the use of pesticides in the production of fibers, poor labor conditions, and animal testing are omnipresent. According to the European Commission, 11.5 million animals were used in the European Union for experimental or scientific purposes in 2011. Worldwide this figure rises to 115 million animals annually (Four Paws International 2013). In the rising tension between "greenwashing" and the use of ethical/environmental commitments that are nothing more than "sheer lip service," the question arises of the role of the consumers with regard to sustainable practices in the cosmetics industry. Are consumers increasingly conscious of the adverse effects of ethical and environmental imbalances? And what effect does this knowledge have on their buying behavior? On the divergent poles of hypocrisy and true commitment, to advance current understanding of sustainability and related links to consumer perception and actual buying behavior related to ethical issues, the aim of this chapter is to provide a comprehensive framework of animal welfare in the personal care industry. Based on existing theoretical and empirical insights it becomes evident that psychological determinants, such as personality traits, empathy, ethical obligation, and self-identity, as well as context-related determinants in terms of one's ethical value perception of products, the trade-off between ethical and conventional products, and an individual's involvement, represent antecedents of ethical consumer behavior, which can be expressed through the avoidance of specific products and brands and/or consumer boycott and buycott towards cosmetics using

N. Hennigs (✉) · E. Karampournioti · K.-P. Wiedmann
Institute of Marketing and Management, Leibniz University of Hannover,
Koenigsworther Platz 1, 30167 Hannover, Germany
e-mail: hennigs@m2.uni-hannover.de

© Springer Science+Business Media Singapore 2016   109
S.S. Muthu and M.A. Gardetti (eds.), *Green Fashion*,
Environmental Footprints and Eco-design of Products and Processes,
DOI 10.1007/978-981-10-0111-6_5

animal-tested ingredients. Our concept provides a useful instrument for both academics and managers as a basis to create and market successfully cosmetics that represent ethical and environmental excellence.

Keywords Animal welfare · Ethical and environmental practices · Global beauty care industry

1 Introduction

> The greatness of a nation and its moral progress can be judged by the way its animals are treated.
>
> Mahatma Gandhi

Driven by rising consumer incomes, changing lifestyles, and a higher demand for luxury products, especially cosmetics, the global beauty care industry is forecast to reach an estimated $265 billion in 2017 (Lucintel 2012). Due to an increased consumer awareness concerning natural and organic products in combination with rising concerns for health safety, the global demand for organic personal care products—the fastest growing segment of the global personal care industry—was noted at $8.4 billion in 2013 and is expected to reach $15.7 billion by 2020 (Transparency Market Research 2015). Apart from consumer's awareness about harmful substances, consumers show rising concerns for animal testing of ingredients and/or finished products. As a consequence, as outlined above, the personal care industry has started to offer organic products without the use of pesticides, synthetic chemicals, and animal testing. However, even if companies commit themselves to advancing good social, environmental, and animal-welfare practices in their business operations, still water pollution, the use of pesticides in the production of fibers, poor labor conditions, and animal testing are omnipresent.

In the rising tension between "greenwashing" and the use of ethical or environmental commitments that are nothing more than "sheer lip service," the question arises of the role of consumers with regard to sustainable practices in the cosmetics industry. Are consumers increasingly conscious of the adverse effects of ethical and environmental imbalances in this market? And what effect does this knowledge have on their buying behavior? On the divergent poles of hypocrisy and true commitment, to advance current understanding of sustainability and related links to consumer perception and actual buying behavior related to ethical issues, the aim of our chapter is to provide a comprehensive framework of animal welfare in the global beauty care industry. To reach this objective, the chapter is structured as follows: first, the theoretical background is provided in the next section, which addresses ethical and environmental consumerism in general and ethical issues in the global beauty care industry in particular. Based on these specifications, a conceptual model of antecedents and outcomes of ethical consumption is derived. Particularly, the framework considers a combination of personality factors and context-related factors as antecedents of brand avoidance and consumer boycott/buycott behavior. Finally, the chapter closes with a discussion of possible future research approaches and managerial implications as opportunities to develop appropriate marketing strategies and adequately respond to their customers' needs and values. In sum, our concept provides a useful instrument for both academics and managers as a basis to create and market successfully personal care goods that represent ethical and environmental excellence.

2 Theoretical Background

2.1 Ethical and Environmental Consumerism

During the last decades, ethical and environmental consumerism has moved from a niche market to a mainstream phenomenon in contemporary consumer culture (Doane 2001; Low and Davenport 2007). The twenty-first century is perceived to be a time of the emancipation of the ethical consumer (e.g., Nicholls 2002), who is concerned about a broad spectrum of issues ranging from the environment and animal welfare to societal concerns, including human rights (Mintel 1994). Strongly related to consumer awareness of conditions in developing countries and the fact that his or her own purchases are connected to social issues (De Pelsmacker and Janssens 2007), the ethical buyer is "shopping for a better world" (Low and Davenport 2007, p. 336) and demands that products are not only friendly to the environment but also to the people who produce them (Rosenbaum 1993). In this context, ethical consumption is related to the purchase of products that concern a certain ethical issue (e.g., human rights, labor conditions, animal well-being, or the environment), the boycott of companies involved in unethical practices, or post-consumption behavior, such as recycling (Jackson 2006; Newholm and Shaw 2007). Consequently, marketing managers in all industries have realized the importance of customer ethics and values and how meeting ethical demands is critical if they wish to gain a competitive advantage (Browne et al. 2000).

2.2 Ethical Issues in the Global Beauty Care Industry

Environmental concern, consumer health orientation, and lifestyle changes have led to a rising demand for green cosmetics and beauty care products without animal testing and harmful substances such as pesticides and synthetic chemicals (Cervellon et al. 2010; Diamantopoulos et al. 2003; Manaktola and Jauhari 2007; Paladino 2006; Papadopoulos et al. 2009; Peter and Olson 2009; Prothero and McDonagh 1992; Pudaruth et al. 2015; Tsakiridou et al. 2008; Zanoli and Naspetti 2002). Prominent brands such as *Aveda, Bare Escentuals, Burt's Bees, Kiehl's, Origins, and The Body Shop* have incorporated the emergence of ethical and environmental consumerism in their business activities by ensuring high environmental standards with emphasis on natural and organic ingredients and animal welfare.

With special focus on animal testing, public resistance to the use of animal studies in the development of cosmetics created the market for products labeled as "animal-free cosmetics" and "non-animal-tested" (see Fig. 1).

Due to the fact that consumers perceived animal tests to be no longer legitimate and referring to a resulting testing and marketing ban in the European Union, the cosmetics industry has to find suitable replacements for animals in cosmetic testing. In particular, cosmetic testing on animals refers to the test of finished products, individual ingredients, and the combination of ingredients on animals. However, even if some cosmetic companies use the claim "not tested on animals", this can be misleading: an ingredient that was once tested and proved to be not harmful can be included in a new product without further tests. Therefore, "non-animal tested" often means "previously animal tested," a fact most consumers are unable to realize (McNeal 2005).

Fig. 1 The rise of cruelty-free cosmetics. (*Sources* http://www.nzavs.org.nz/nz-passes-cosmetics-animal-testing-ban; http://mumbrella.com.au/animal-rights-group-behind-banned-graphic-violence-ad-parts-ways-with-agency-work-deemed-not-shocking-enough-107638; http://action.peta.org.uk/ea-action/action?ea.client.id=5&ea.campaign.id=15529; http://www.picturequotes.com/thank-you-eu-for-banning-cruel-cosmetics-quote-25128; http://www.thebodyshop.com/values/EU_Against_Animal_Testing.aspx; http://www.leapingbunny.org/downloads; http://www.peta.org/living/beauty/beauty-without-bunnies/; http://www.novenamaternity.com/certifications/; http://www.tierschutzbund.de/information/hintergrund/tierversuche/kosmetik.html)

In fact, even though testing cosmetics on animals is banned in many countries, it is still omnipresent in the personal care industry. In China, animal testing is mandatory; in the United States, animal testing for cosmetics products or ingredients is not required, but "animal testing by manufacturers seeking to market new products is often necessary to establish product safety" (FDA 2000). Cosmetic products that have not been adequately tested for safety must have a warning statement on the front label "WARNING—The safety of this product has not been determined" (FDA 2000). Even though experiments on animals are cast in a negative light (see Fig. 2) and governmental regulations try to reduce their implementation, it is estimated that approximately 115 million animals are used for laboratory experiments worldwide (HIS 2012). However, there is criticism that data are not fully covered. In the United States, for example, nearly 90 % of used animals are not covered by official statistics so that the number of 834,453 reported cases for 2014 (USDA 2015) may be far higher than estimated (HIS 2012).

Fig. 2 Campaigns against animal testing. (*Sources* http://mumbrella.com.au/animal-rights-group-behind-banned-graphic-violence-ad-parts-ways-with-agency-work-deemed-not-shocking-enough-107638; http://www.peta2.com/heroes/noah-cyrus-dissection-kills/; https://www.pinterest.com/pin/224617100138736978/; http://de.adforum.com/creative-work/ad/player/34454904; http://blog.peta.org.uk/wp-content/uploads/2011/07/PETA.pdf; http://www.sanjeev.net/printads/l/lavera-the-price-of-beauty-695.html; http://www.peta.org/features/dave-navarro-cruelty-free/)

3 Conceptual Model

Ethical blunders of companies such as the acceptance and conducting of experiments on animals result in significant impacts on consumer behavior. Therefore, and against the backdrop of the challenges as discussed above, companies need to gain an understanding of underlying determinants and possible outcomes of ethical consumption. Because components of consumption behavior are not exclusively limited to the purchase or consumption of specific products or brands but are focusing on resistant or "against consumption" behavior as well (Lee et al. 2009; Varman and Belk 2009), the present work focuses on behavioral outcomes such as boycott/buycott and brand avoidance.

To reach this objective, psychological consumer traits as well as context-related issues are summarized within a shared model. For a structured and comprehensive overview, the conceptual framework in Fig. 3 considers a combination of *personality factors* (i.e., personality traits, empathy, ethical obligation, and self-identity) and *context-related factors* (i.e., ethical value perception, involvement, and the trade-off between ethical and conventional products) as antecedent's of *brand avoidance and consumer boycott and buycott* towards cosmetics using animal-tested ingredients.

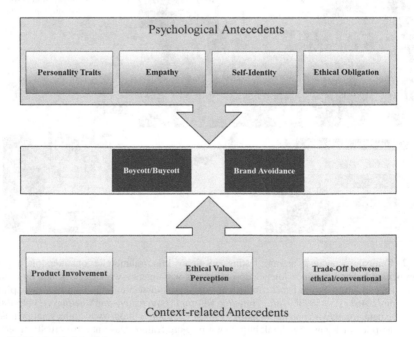

Fig. 3 Conceptual model

3.1 Personality Factors

- *Personality Traits*: In an attempt to explain consumer behavior in general and ethical consumption in particular, consumers' personality traits are often related to purchases or nonpurchases of specific products or brands. Therefore, consuming in a particular manner is largely determined by personality characteristics and ethical decision making grounded on personal characteristics of individuals (Grubb and Grathwohl 1967, Hunt and Vitell 1986, 1992; Ferrell and Gresham 1985). In spite of the fact that "there appears to be as many definitions of personality as there are authors" (Pervin 1990, p. 3), the term is subject to several definitions and understandings of its meaning. Based on the assumption of a temporal stability (Peck and Whitlow 1975), personality represents "generalized patterns of response or modes of coping with the world ..." (Kassarjian 1971, p. 409). In contrast, Triandis (2001) conceptualized personality as "a configuration of cognitions, emotions, and habits activated when situations stimulate their expression" (p. 908) and reveals that one's personality undergoes continuous changes and is to a high degree influenced by the external environment.

  Some studies have examined the role of personality traits for attitudes towards animal testing in general. Broida et al. (1993) reveal that extraverted and conservative personality traits are positively correlated with animal testing. Furthermore, agreeableness, extraversion, and openness, belonging to the big five personality traits, are consistently and logically related to animal welfare and have a strong predictive power for negative attitudes towards animal testing (Furnham et al. 2003). Additionally, Goldsmith et al. (2006) focused on animal-tested cosmetics and discovered that higher levels of anticonformity were associated with opposition to animal testing as well. Hence, we propose that the receiveability of ethical dilemmas, such as practices against animal welfare, and the willingness of individuals to act against them, largely depends on personality factors (Vitell and Muncy 1992; Munch et al. 1991).

- *Empathy*: At its core, empathy refers "in various ways to the experiencing of another's affective or psychological state and has both affective and cognitive components" (Zahn-Waxler and Radke-Yarrow 1990, p. 108). Although the cognitive component "entails understanding or identifying with another individual's response" (McPhedran 2009, p. 1) and is therefore sometimes labeled as "perspective taking," the affective component puts emphasis on an individual's emotional response "that is congruent with and stems from the apprehension of another's emotional state or condition" (Zahn-Waxler and Radke-Yarrow 1990, p. 108) and involves sharing (empathic concern) as well as reacting to (personal distress) emotional experiences (Davis 1980; McPhedran 2009; Signal and Taylor 2007; Eisenberg and Strayer 1987). The ability to empathize is not limited to human-to-human interactions but comprises those to animals as well (Apostol et al. 2013). Due to the reason that individuals capable of empathy are able to experience the consequences of their actions on others, it is more likely that harmful behaviour will be avoided (McPhedran 2009) and that consumption behavior will be adjusted to solve ethical issues regarding the execution of animal experiments for the production of cosmetics.

- *Ethical Obligation*: A sense of ethical or moral obligation constitutes a driver for decision making in the context of ethical consumerism (Shaw and Shiu 2002) and represents "an individual's ethical rules, which reflect their personal beliefs about right or wrong" (Shaw et al. 2000, p. 882). Increasing numbers of consumers intend "to make certain consumption choices due to personal and moral beliefs" (Crane and Matten 2004, p. 341). Considering ethical and moral issues when choosing products and services is "used to cover matters of conscience such as animal welfare" (Cowe and Williams 2000, p. 4) and because for some consumers "it is the right thing to do" (Carrigan et al. 2004, p. 402). Thus, it is expected that consumers who perceive the obligation to show moral concern for animals may avoid products/cosmetics with animal-tested ingredients.
- *Self-Identity*: Consumerism is not only a way to satisfy physiological needs, because consumers try to express themselves and to define their identities through the use or the avoidance of specific products and brands (Carrigan et al. 2004; Aaker 1999; Solomon 1983; Hogg et al. 2000). Self-identity refers to "relatively enduring characteristics that people ascribe to themselves," and is synonymously used with self-concept or self-perception (Sparks and Guthrie 1998, p. 1396; Hustvedt and Dickson 2009). Despite an individual's real self, imaginary selves exist, often classified in desired (positive) and undesired (negative) selves (Ogilvie 1987; Markus and Nurius 1986). To avoid identification with the undesired self, consumers tend to avoid specific products or services and the representation of one's desired self-concept can be promoted through consumption (Banister and Hogg 2004; Wright et al. 1992; Freitas et al. 1997; Karanika and Hogg 2010). Hence, if ethical issues, as represented by animal welfare concerns, have become an important part of an individual's self-identity, consumption choices and antichoices will be adjusted based on them (Shaw et al. 2000).

3.2 Context-Related Factors

- *Product Involvement*: Involvement can be seen as "a person's perceived relevance of the object based on inherent needs, values, and interests" (Zaichkowsky 1985, p. 342) and is understood as an internal "motivational state" (Mittal 1989) or unobservable "state of motivation" (Rothschild 1984) that indicates the intensity of arousal or interest. Further research demonstrated that this construct has strong predictive power for consumers' behavior (Celsi and Olson 1988; Zaichkowsky 1985, Dholakia 2001). In the specific case of ethical consumerism, involvement is not limited to specific products or brands, but is additionally related to their ethical augmentation (Crane 2001; Bezençon and Blili 2010). Accordingly, a high level of involvement in ethical issues in general and in animal welfare issues in particular influences the consumers' search for ethical information of specific products or brands as well as their receptivity to them which consequently affects behavioral intentions (Celsi and Olson 1988; Greenwald and Leavitt 1984; Zaichkowsky 1985).

- *Ethical Value Perception*: The value of ethical consumption as perceived by consumers and subsequently the importance of meeting ethical demands has a considerable impact on the achievement of competitive advantages (Browne et al. 2000). To investigate the question of what really adds value, it is essential to consider the multidimensionality of the customer's perceived value based on "consumer's overall assessment of the utility of a product (or service) based on perceptions of what is received and what is given" (Zeithaml 1988, p. 14). According to previous research on customer-perceived value by Sweeney and Soutar (2001) as well as by Smith and Colgate (2007), well-known consumption values can be commonly divided into the four types *economic, functional, affective, and social*. The *economic value* refers to direct monetary aspects of the product expressed in dollars and cents that one is willing to spend to obtain a product (Ahtola 1984; Monroe and Krishnan 1985). The basic utilities and benefits of the product such as quality, uniqueness, usability, reliability, and durability are part of the *functional value* (Sheth et al. 1991). The *affective value* describes the perceived subjective utility attained through the consumption of a product and the related arousal of feelings and affective states to ethical consumers who assign high importance to aspects such as altruism, equality, and peace (Littrell and Dickson 1999). Recently, the *social dimension* of customer-perceived value mentions the desire of ethical consumers to try to impress and to meet the expectations of their social group and to influence the perception of others' judgment of one's own behavior. However, even if individuals lack intrinsic value to consume ethically, they would still behave ethically through the pressure of social norms (Starr 2009). With reference to cosmetics, it is expected that consumers who have a high value perception of ethical product characteristics and business practices are less willing to purchase products/cosmetics tested on animals.

- *Trade-Off Between Ethical and Conventional Products*: By means of their consumption choice, consumers try to satisfy their immediate consumption needs and conscience as well (Ehrich and Irwin 2005). Furthermore, "consumers will probably not sacrifice aspects of product performance for ethical considerations alone" (Auger et al. 2008, p. 190). Hence, the individual choice decision between ethical and conventional cosmetics is influenced by the trade-off between a given product's ethical features such as consideration of animal rights and avoidance of animal experiments, its functional performance (Luchs et al. 2007), and the price of cosmetics free from animal testing (Auger et al. 2010).

3.3 Related Outcomes

- *Boycott*: In a boycott, which constitutes a typical expression of ethical consumption, consumers desist from buying certain products or brands. Their resistance is based upon an ideological displeasure with an organization (Friedman 1985; Hirschmann 1970) and represents "an attempt by one or more parties to achieve certain objectives by urging individual consumers to refrain from making selected purchases in the marketplace" (Friedman 1985, p. 97). Accordingly, boycott behavior intends "to benefit one or more people other than oneself behaviors such as helping, comforting, sharing, and cooperation" (Batson 1998, p. 282) and is used to punish undesirable business behavior (Hofmann and Hutter 2012) and to protest against unfair company practices of social, ethical, moral, or environmental nature (Delacote 2006; Diermeier and van Mieghem 2005). Whereas boycotts represent a useful means to punish companies for their misbehavior, *buycotts* follow the opposite approach and serve as a reward system for past good deeds (Hawkins 2010) which "attempt to induce shoppers to buy the products or services of selected companies in order to reward them for behavior which is consistent with the goals of the activists" (Friedman 1996, p. 440). Both forms of activism can occur simultaneously; this is partly because consumers participate with a higher probability to boycotts if the targeted product has satisfactory substitutes (Sen et al. 2001). Although consumers may boycott cosmetics accepting experiments on animals, concurrent preferment of those products and brands might occur, which refrain from such cruel business practices.
- *Brand Avoidance*: The targeted rejection of a brand can be defined as brand avoidance. According to Lee et al. (2009) reasons for the avoidance can be the perceived incongruence between the brand and the customer's desired or actual self-concept (*identity avoidance*; Englis and Soloman 1997; Grubb and Grathwohl 1967; Hogg and Banister 2001; Sirgy 1982), dissatisfaction through negative brand consumption experiences (*experiential avoidance*; Folkes 1984; Oliver 1980) and the "belief that it is a moral duty to avoid certain brands" (Lee et al. 2009, p. 7) due to the existence of an ideological incompatibility between consumer and brand (*moral avoidance*). Boycotters of animal-tested cosmetics provide the probability to rebuild the relationship if certain conditions are met (Hirschman 1970), however, the avoidance of a brand offers no guarantee for a possible reconstruction (Lee et al. 2009).

Based on the holistic understanding of the psychological and context-related determinants as well as related outcomes as described above, our comprehensive framework is a basis to gain a structured understanding of underlying determinants and possible outcomes of consumer behavior in the global beauty care industry. However, it has to be stated that consumers often tend to act differently depending on whom they are interacting with and the situation they are in. In addition, taking into account that consumers' positive attitudes towards green cosmetics and against animal testing do not necessarily transfer into consumption behavior, an attitude–behavior gap often exists between consumer claims and actual behavior (Carrigan and Attalla 2001; Bhattacharya and Sen 2004; Öberseder et al. 2011). Therefore, even if consumers report positive perceptions of the psychological and context-related determinants as proposed in our model, it is not possible to predict how consumers will behave in a real purchase situation and if the attitude–behavior gap can be minimized or overcome.

In addition to these considerations, apart from the positive attitude towards green cosmetics and animal rights protection, existing studies give evidence to concentrate on the "dark side" of the consumer personality as well. As can be seen in previous research, *Machiavellianism, narcissism, and psychopathy*—collectively known as the *Dark Triad* of personality traits—play an increasingly important role in society. This importance becomes particularly evident through the overemphasis of the self and the self-promotion through social media as well as through the increasing research effort concerning the workplace behavior of "snakes in suits" and "bad bosses" (for a detailed overview see Furnham et al. 2013; Garcia and Sikström 2014; Buckels et al. 2014). With special focus on animals, Kavanagh et al. (2013) detected that individuals with high levels of *Dark Triad traits* demonstrated less positive attitudes towards animals and have even practiced violence against them.

Characteristics of the *Dark Triad* include "entitlement, superiority, dominance (i.e., narcissism), glib social charm, manipulativeness (i.e., Machiavellianism), callous social attitudes, impulsivity, and interpersonal antagonism (i.e., psychopathy)" (Jonason et al. 2015, p. 6). Even if the individual characteristics have different origins, "all three entail a socially malevolent character with behavior tendencies toward self-promotion, emotional coldness, duplicity, and aggressiveness" (Paulhus and Williams 2002, p. 557). Because individuals with high levels of Dark Triad scales, value the "self" over "other", these traits are often associated with reduced or dysfunctional morality (Campbell et al. 2009; Glenn et al. 2009) which could have a considerably high impact on the perception and evaluation of ethical and moral issues and subsequently on their consumption behavior in general and on the cosmetics industry in particular.

Taken as a whole, based on the preceding insights and related discussion, several implications for further research and managerial practice can be drawn as presented in the concluding remarks.

4 Conclusion

Confronted with criticisms on irresponsible business activities such as the use of animal testing, chemical pollution, unethical sourcing, and unsustainable ingredients, the global beauty care industry has realized the importance of ethical and environmental business practices. The adoption of corporate social responsibility activities is reflected in efficient use of energy and water, avoidance of animal testing and unethical ingredients, reduction of packaging, use of environmentally friendly packaging material, distribution via ethical supply chains, and fair trade (Organic Monitor 2010). From a consumer perspective, rising concerns, associated with health-related issues and ethical or environmental qualities of the products they buy, have led to an increased demand for natural and organic products. With special focus on the global beauty care industry, consumers increasingly value organic products that are free from harmful substances and animal testing.

Referring to the antecedents and outcomes of consumer perception and behavior in the context of green cosmetics, the aim of this chapter was to present a holistic framework of psychological consumer traits as well as context-related issues and related outcomes. Our model can be seen as a useful basis to create and market personal care goods successfully that represent ethical and environmental excellence.

Focusing on future research, the determinants included in the framework have to be empirically tested with reference to different consumer groups and product-specific contexts. It is expected that the relative impact of the antecedents on actual consumption behavior differs in consideration of the variety of cultures across the world as well as different consumer lifestyles and consumption patterns within national borders.

Based on a better knowledge of relevant drivers and outcomes of ethical consumption, marketers in the global beauty care industry can compare the core values expressed by their brand and compare them to the individual aspiration level of their actual and potential consumers to develop appropriate marketing strategies and adequately respond to their customers' needs and values. To verify that the commitment to ethical values and animal welfare is more than a clever promotional gimmick and to refute accusations of greenwashing, ethical orientation has to become part of the corporate culture and business model. Each management decision has to be reflected from the ethical perspective and the responsibility that consumers expect inherent in the multifaceted product attributes. To separate hypocrisy and true commitment clearly, companies in the global beauty care industry have to redefine their products and production processes, examine the supply chains, and translate social and environmental strategies into operational practices. Instead of considering ethical obligations as a threat to corporate profits, incorporating ethical and environmental excellence has to be regarded as a successful business opportunity in a promising way to reconcile financial, ethical, and ecological values:

> Our task must be to free ourselves ... by widening our circle of compassion, to embrace all living creatures in the whole of nature and its beauty.
>
> Albert Einstein

References

Aaker JL (1999) The malleable self: the role of self-expression in persuasion. Mark Res 36:45–57
Ahtola OT (1984) Price as a 'give' component in an exchange theoretic multicomponent model. Adv Consum Res 11(1):623–636

Apostol L, Rebega OL, Miclea M (2013) Psychological and socio-demographic predictors of attitudes toward animals. Procedia Soc Behav Sci 78:521–525

Auger P, Devinney TM, Louviere JJ et al (2008) Do social product features have value to consumers? Int J Res Mark 25(3):183–191

Auger P, Devinney TM, Louviere JJ et al (2010) The importance of social product attributes in consumer purchasing decisions: a multi-country comparative study. Int Bus Rev 19(2):140–159

Banister EN, Hogg MK (2004) Negative symbolic consumption and consumers' drive for self-esteem. Eur J Mark 38(7):850–868

Batson CD (1998) Altruism and prosocial behavior. In: Lindzey, G, Gilbert, D, Fiske, ST (eds) The handbook of social psychology. Oxford University Press, Oxford, pp 282–301

Bezençon V, Blili S (2010) Ethical products and consumer involvement: what's new? Eur J Mark 44(9/10):1305–1321

Bhattacharya CB, Sen S (2004) Doing better at doing good: when, why, and how consumers respond to corporate social initiatives. Calif Manage Rev 47(1):9–24

Broida J, Tingley L, Kimball R et al (1993) Personality differences between pro-and antivivisectionists. Soc Anim 1(2):129–144

Browne A, Harris P, Hofney-Collins A et al (2000) Organic production and ethical trade: definition, practice and links. Food Policy 25(1):69–89

Buckels EE, Trapnell PD, Paulhus DL (2014) Trolls just want to have fun. Pers Indiv Differ 67:97–102

Campbell J, Schermer JA, Villani VC et al (2009) A behavioral genetic study of the Dark Triad of personality and moral development. Twin Res Hum Genet 12(02):132–136

Carrigan M, Attalla A (2001) The myth of the ethical consumer—do ethics matter in purchase behavior? J Consum Mark 18(7):560–578

Carrigan M, Szmigin I, Wright J (2004) Shopping for a better world? An interpretive study of the potential for ethical consumption within the older market. J Consum Mark 21(6):401–417

Celsi RL, Olson JC (1988) The role of involvement in attention and comprehension processes. J Consum Res 15(2):210–224

Cervellon MC, Hjerth H, Richard S et al (2010) Green in fashion? An exploratory study of national differences in consumers concern for eco-fashion. In: Proceedings of 9th international marketing trends conference, Venice, 20–21 Jan 2010

Cowe R, Williams S (2000) Who are the ethical consumers. Co-operative Bank, Manchester

Crane A (2001) Unpacking the ethical product. J Bus Ethics 30(4):361–373

Crane A, Matten D (2004) Business ethics: a European perspective: managing corporate citizenship and sustainability in the age of globalization. Oxford University Press, Oxford

Davis MH (1980) A multidimensional approach to individual differences in empathy. JSAS Catalogue Sel Doc Psychol 10:85

De Pelsmacker P, Janssens W (2007) A model for fair trade buying behaviour: the role of perceived quantity and quality of information and of product-specific attitudes. J Bus Ethics 75(4):361–380

Delacote P (2006) Are consumer boycotts effective? Paper presented at the 6th IDEI-LERNA conference on environmental resource economics—environment, finance and corporate behavior, Toulouse, May 2007

Dholakia UM (2001) A motivational process model of product involvement and consumer risk perception. Eur J Mark 35(11/12):1340–1362

Diamantopoulos A, Schlegelmilchh BB, Sinkovics RR et al (2003) Can sociodemographics still play a role in profiling green consumers? A review of the evidence and an empiricalinvestigation. J Bus Rev 56(6):465–480

Diermeier D, Van Mieghem J (2005) A stochastic model of consumer boycotts. Department of Managerial Economics and Decision Sciences (MEDS), Kellogg School of Management, Northwestern University, Evanston

Doane D (2001) Taking flight: the rapid growth of ethical consumerism, the ethical purchasing index 2001. New Economics Foundation, London

Ehrich KR, Irwin JR (2005) Willful ignorance in the request for product attribute information. J Mark Res 42(3):266–277

Eisenberg N, Strayer J (1987) Critical issues in the study of empathy. In: Eisenberg N, Strayer J (eds) Empathy and its development. Cambridge University Press, Cambridge, pp 3–13

Englis BG, Soloman MR (1997) Special session summary: I am not therefore, I am: the role of avoidance products in shaping consumer behavior. Adv Consum Res 24:61–63

FDA (2000) Animal testing, U.S. Food and Drug Administration, Office of Cosmetics and Colors Factsheet. Center for Food Safety and Applied Nutrition, 14 Mar 1995, Revised 24 Feb 2000

Ferrell OG, Gresham L (1985) A contingency frame work for understanding ethical decision making in marketing. J Mark 49(3):87–96

Folkes VS (1984) Consumer reactions to product failure: an attributional approach. J Consum Res 10(4):398–409

Four Paws International (2013) World day for laboratory animals, 24 Apr 2013. http://www.four-paws.org.uk/news-press/news/world-day-for-laboratory-animals-24thapril-2013-/. Accessed 11 June 2015

Freitas A, Kaiser S, Chandler J et al (1997) Appearance management as border construction: least favourite clothing, group, distancing, and identity…Not! Sociol Inq 67(3):323–335

Friedman M (1985) Consumer boycotts in the United States, 1970–1980: contemporary events in historical perspective. J Consum Aff 19(1):96–117

Friedman M (1996) A positive approach to organized consumer action: The "buycott" as an alternative to the boycott. J Consum Psychol 19(4):439–451

Furnham A, McManus C, Scott D (2003) Personality, empathy and attitudes to animal welfare. Anthrozoos 16(2):135–146

Furnham A, Richards SC, Paulhus DL (2013) The Dark Triad of personality: a 10 year review. Soc Personal Psychol Compass 7(3):199–216

Garcia D, Sikström S (2014) The dark side of Facebook: semantic representations of status updates predict the Dark Triad of personality. Pers Indiv Differ 67(September):92–96

Glenn AL, Iyer R, Graham J et al (2009) Are all types of morality compromised in psychopathy? J Pers Disord 23(4):384–398

Goldsmith RE, Clark RA, Lafferty B (2006) Intention to oppose animal research: the role of individual differences in nonconformity. Soc Behav Personal 34(8):955–964

Greenwald AG, Leavitt C (1984) Audience involvement in advertising: four levels. J Consum Res 11(1):581–592

Grubb EL, Grathwohl HL (1967) Consumer self-concept, symbolism and market behavior: a theoretical approach. J Mark 31(4):22–27

Hawkins RA (2010) Boycotts, buycotts and consumer activism in a global context: an overview. MOH 5(2):123–143

Hirschmann AO (1970) Exit, Voice, and Loyalty: Response to Decline in Firms, Organizations, and States. Harvard University Press

Hoffmann S, Hutter K (2012) Carrotmob as a new form of ethical consumption. The nature of the concept and avenues for future research. J Consum Policy 35(2):215–236

Hogg MK, Cox AJ, Keeling K (2000) The impact of self monitoring on image congruence and product/brand evaluation. Eur J Mark 34(5/6):641–666

Hogg MK, Banister EN (2001) Dislikes, distastes and the undesired self: conceptualising and exploring the role of the undesired end state in consumer experience. J Mark Manage 17(1–2):73–104

HIS (2012) Animal use statistics. Humane Society International. http://www.hsi.org/campaigns/end_animal_testing/facts/statistics.html. Accessed 19 July 2015

Hunt Shelby D, Vitell Scott M (1992) The general theory of marketing ethics: A retrospective and revision. Ethics in marketing. Irwin, Homewood, pp 775–784

Hunt SD, Vitell S (1986) A general theory of marketing ethics. J Macromark 6(1):5–16

Hustvedt G, Dickson MA (2009) Consumer likelihood of purchasing organic cotton apparel: influence of attitudes and self-identity. J Fashion Mark Manage: Int J 13(1):49–65

Jackson T (2006) Challenges for Sustainable Consumption Policy. In: Jackson T (ed) The earthscan reader on sustainable consumption. Earthscan, London, pp 109–128

Jonason PK, Baughman HM, Carter GL et al (2015) Dorian without his portrait: The psychological, social, and physical health costs of the Dark Triad traits. Pers Indiv Differ 78:5–13

Karanika K, Hogg MK (2010) The interrelationship between desired and undesired selves and consumption: the case of Greek female consumers' experiences. J Mark Manage 26(11–12):1091–1111

Kassarjian HH (1971) Personality and consumer behavior: a review. J Mark Res 8(4):409–418

Kavanagh PS, Signal TD, Taylor N (2013) The Dark Triad and animal cruelty: Dark personalities. Pers Indiv Differ 55(6):666–670

Lee MS, Motion J, Conroy D (2009) Anti-consumption and brand avoidance. J Bus Rev 62(2):169–180

Littrell MA, Dickson MA (1999) Social responsibility in the global market: fair trade of cultural products. Sage Publications, New York

Low W, Davenport E (2007) To boldly go. Exploring ethical spaces to re-politicise ethical consumption and fair trade. JCB 6(5):336–348

Luchs M, Naylor RW, Irwin JR et al (2007) Is there an expected trade-off between a product's ethical value and its effectiveness? Exposing latent intuitions about ethical products. J Int Mark 11(2):101–111

Lucintel (2012) Global Beauty Care Products Industry 2012-2017: trend, profit, and forecast analysis, Sep 2012

Manaktola K, Jauhari V (2007) Exploring consumer attitude and behaviour towards greenpractices in the lodging industry in India. Int J Contemp Hosp M 19(5):364–377

Markus H, Nurius P (1986) Possible selves. Am Psychol 41(9):954–969

McNeal KR (2005) Death: the price of beauty: Animal Testing and the Cosmetics Industry, American Bar Association Section of Environment, Energy and Resources, Law Student Division, Spring 2005. http://www.abanet.org/environ/committees/lawstudents/pdf/mcneal. pdf. Accesses 22 June 2015

McPhedran S (2009) A review of the evidence for associations between empathy, violence, and animal cruelty. Aggress Violent Beh 14(1):1–4

Mintel (1994) The green consumer. Mintel Research, London

Mittal B (1989) Measuring purchase-decision involvement. Psychol Market 6(2):147–162

Monroe KB, Krishnan R (1985) The effect of price on subjective product evaluations. In: Jacoby J, Olson J (eds) The perception of merchandise and store quality. Lexington Books, Lexington, pp 209–232

Munch JM, Albanese PJ, Mayo MA et al (1991) The role of personality and moral development in consumers 'Ethical Decision Making'. In: Proceedings of American marketing summer educator's conference, pp 299–308

Newholm T, Shaw D (2007) Studying the ethical consumer: a review of research. J Consum Behav 6(5):253–270

Nicholls AJ (2002) Strategic options in fair trade retailing. Int J Retail Distrib Manage 30(1):6–17

Öberseder M, Schlegelmilch B, Gruber V (2011) Why don't consumers care about CSR?: a qualitative study exploring the role of CSR in consumption decisions. J Bus Ethics 104(4):449–460

Ogilvie DM (1987) The undesired self: a neglected variable in personality research. J Pers Soc Psychol 52(2):379–385

Oliver RL (1980) A cognitive model of the antecedents and consequences of satisfaction decisions. J Mark Res 17(4):460–469

Organic Monitor (2010) CSR & Sustainability: How the Beauty Industry is Cleaning up, 18 May 2010. http://www.organicmonitor.com/r1805.htm. Accessed 11 June 2015

Paladino A (2006) Understanding the green consumerism: an empirical analysis. J Consum Behav 4(1):69–102

Papadopoulos I, Karagouni G, Trigkas M et al (2009) Green marketing: the case of timber certification, coming from sustainable forests management, promotion. In: Annual international
Euro med conference proceedings, vol 2. The Research Business Institute, Salemo

Paulhus DL, Williams KM (2002) The dark triad of personality: Narcissism, Machiavellianism,
and psychopathy. J Res Pers 36(6):556–563

Peck D, Whitlow D (1975) Approaches to personality theory. Methuen, London

Pervin LA (1990) A brief history of modern personality theory. In: Pervin LA (ed) Handbook of
personality theory and research. The Guilford Press, New York, pp 3–18

Peter JP, Olson JC (2009) Consumer behaviour and marketing strategy. McGraw Hill, New York

Prothero A, McDonagh P (1992) Producing environmentally acceptable cosmetics? The impact
of environmentalism on the United Kingdom cosmetics and toiletries industry. J Mark
Manage 8(2):147–166

Rosenbaum M (1993) Trading Standards: Will customers now shop for fair play?: Group plans
to flag products that are Third World-friendly. The Independent. http://www.independent.
co.uk/news/business/trading-standards-will-customer-now-shop-for-fair-play-group-plans-
to-flag-products-that-are-third-worldfriendly-1489932.html. Accessed 13 July 2015

Rothschild ML (1984) Perspectives on involvement: current problems and future directions. Adv
Consum Res 11(1):216–217

Sen S, Gürhan-Canli Z, Morwitz V (2001) Withholding consumption: a social dilemma perspective on consumer boycotts. J Cons Res 28(3):399–417

Pudaruth S, Juwaheer TD, Seewoo YD (2015) Gender-based differences in understanding the
purchasing patterns of eco-friendly cosmetics and beauty care products in Mauritius: a study
of female customers. Soc Responsib J 11(1):179–198

Shaw D, Shiu E (2002) An assessment of ethical obligation and self-identity in ethical consumer decision-making: a structural equation modelling approach. Int J Consum Stud
26(4):286–293

Shaw D, Shiu E, Clarke I (2000) The contribution of ethical obligation and self-identity to
the theory of planned behaviour: an exploration of ethical consumers. J Mark Manage
16(8):879–894

Sheth JN, Newman BI, Gross BL (1991) Why we buy what we buy: a theory of consumption
values. J Bus Rev 22(2):159–170

Signal T, Taylor N (2007) Attitude to animals and empathy: comparing animal protection and
general community samples. Anthrozoos 20(2):125–130

Sirgy JM (1982) Self-concept in consumer behavior: a critical review. J Consum Res
9(3):287–300

Smith JB, Colgate M (2007) Customer value creation: a practical framework. J Mark Theory
Pract 15(1):7–23

Solomon MR (1983) The role of products as social stimuli: a symbolic interactionism perspective. J Consum Res 10(3):319–329

Sparks P, Guthrie CA (1998) Self-identity and the theory of planned behavior: a useful addition
or an unhelpful artifice? J Appl Soc Psychol 28(15):1393–1410

Starr MA (2009) The social economics of ethical consumption: theoretical considerations and
empirical evidence. J Socio-Econ 38(6):916–925

Sweeney JC, Soutar GN (2001) Consumer-perceived value: the development of a multiple item
scale. J Retail 77(2):203–220

Transparency Market Research (2015) Organic Personal Care Products Market—Global Industry
Analysis, Size, Share, Growth, Trends and Forecast, 2014–2020. http://www.transparencym
arketresearch.com/organic-personal-care-products.html. Accessed 11 Aug 2015

Triandis HC (2001) Individualism-collectivism and personality. J Pers 69(6):907–924

Tsakiridou E, Boutsouki C, Zotos Y et al (2008) Attitudes and behaviour towards organic products: an exploratory study. Int J Retail Distrib Manag 36(2):158–175

USDA (2015). Annual report animal usage by fiscal year. United States Department of Agriculture Animal and Plant Health Inspection Service. http://www.aphis.usdAgov/animal_welfare/downloads/7023/Animals%20Used%20In%20Research%202014.pdf. Accessed 19 Aug 2015

Varman R, Belk RW (2009) Nationalism and ideology in an anticonsumption movement. J Consum Res 36(4):686–700

Vitell SJ, Muncy J (1992) Consumer ethics: An empirical investigation of factors influencing ethical judgments of the final consumer. J Bus Ethics 11(8):585-597

Wright ND, Claiborne CB, Sirgy MJ (1992) The effects of product symbolism on consumer self-concept. Adv Consum Res 19(1):311 318

Zahn-Waxler C, Radke-Yarrow M (1990) The origins of empathic concern. Motiv Emot 14(2):107–130

Zaichkowsky JL (1985) Measuring the involvement construct. J Consum Res 13(3):341–352

Zanoli R, Naspetti S (2002) Consumer motivations in the purchase of organic food: a means-end approach. Br Food J 104(8):643–653

Zeithaml VA (1988) Consumer perceptions of price, quality, and value: a means-end model and synthesis of evidence. J Mark 52(3):2–22

Sustainable Value Generation Through Post-retail Initiatives: An Exploratory Study of Slow and Fast Fashion Businesses

Rudrajeet Pal

Abstract The dialogue between slow and fast fashion has gained great prominence in recent years particularly in terms of sustainability. In the forward value chain, fast fashion companies are most often considered to be unsustainable whereas the slow fashion brands are comparatively more planet-friendly. However, the discussion on the trade-off between sustainability and "speed of fashion" (classified into slow and fast fashions) in the post-retail segment is still limited. A deeper understanding, however, would not only contribute towards conceptualizing the post-retail initiatives, but would also shed light on how these are differentially undertaken by various types of fashion businesses in terms of generating sustainable value. This study proposes sustainable value generation in terms of closing the material and responsibility loops. It further reveals that the trade-off in post-retail is not as rigid as it is in the forward value chain. However, fast fashion offers the lowest potential to displace the purchase of new clothes to close the material loop whereas the redesign brands offer the highest; moreover fast fashion is less liable to take extended responsibilities compared to the slow fashion brands. It can be concluded that fast fashion is somewhat "stuck in the middle" in comparison to the slow and redesign brands in terms of generating value through closing the loop activities.

Keywords Fast fashion · Slow fashion · Post-retail · Responsibility · Business model · Value

1 Introduction

The dialogue between slow and fast fashion has gained great prominence in recent years, particularly in terms of sustainability (Clark 2008; Fletcher 2010). In the forward value chain, fast fashion companies are often considered to be

R. Pal (✉)
Department of Business Administration and Textile Management,
University of Borås, Borås, Sweden
e-mail: rudrajeet.pal@hb.se

© Springer Science+Business Media Singapore 2016   127
S.S. Muthu and M.A. Gardetti (eds.), *Green Fashion*,
Environmental Footprints and Eco-design of Products and Processes,
DOI 10.1007/978-981-10-0111-6_6

unsustainable due to their increasing favor towards mass production and consumption. Owing to various factors, such as production in bulk quantities, cost-intensity and low quality, short product lifetime, and so on, fast fashion retailers are criticized for contributing towards a cheap, throw-away society plagued by surpluses (Birtwistle and Moore 2007). On the other hand, the slow fashion brands are considered comparatively more planet-friendly owing to their diverse practices of supporting local manufacturing, durable or timeless product designs, reuse activities, slow consumption, and the like (Fletcher 2010; Fletcher and Grose 2012).

In this growing sustainability paradigm, however, the focus towards value chains has shifted from being linear to circular. Various efforts are made by organizations in order to bring back the disposed-of products and close the forward chain material loop by using reverse logistics activities such as reuse, refurbish, repair, remanufacture, recycle, or redesign (Jayaraman and Luo 2007). This circularity of value chains has gained momentum in the fashion and apparel industry as well, as fashion businesses have started engaging with various post-retail initiatives, including reuse, recycle, remanufacture, and so on. For example, many fashion brands, including Patagonia and Nudie Jeans have started schemes to take back their old products from wearers and sell them as second-hand, either "as is" or after some redesign. These sales are conducted either through their own stores or via online marketplaces (such as e-Bay; Ekvall et al. 2014). Large retailers including H&M and PUMA have similarly entered into collaborations, but with global sorting partners to develop efficient take-back schemes. Some of the retailers such as Marks and Spencer and Levi's have initiated similar initiatives in collaboration with charities. On the other hand, many smaller brands have started redesigning the used clothes for a longer life thus fitting into the slow fashion movement philosophy (Gardetti and Torres 2013). For example, Studio Re:design is a regional initiative in West Sweden aimed at upcycling textile and clothing wastes and leftovers from the Red Cross using slow craft principles (Studio Re:design 2014). Under this initiative many small design brands have taken part in the redesign collective to aim for attaining a circular economy by redesigning wastes. In connection many small redesign brands have eventually evolved combining traditional tailoring, repairing, and customization services to create new garments from used ones that have been donated by customers, producers, or charities. These aim at rendering sustainable solutions for handling the garments' end of life (Ekvall et al. 2014). In this post-retail market context, companies address sustainable value generation in various ways, along 5-R: reuse, reduce, recycle, redesign, and reimagine (Ho and Choi 2012) that is crucial to attain:

1. Higher resource efficiency through higher end-of-use (EOU) recovery
2. Lower virgin material consumption
3. Higher responsibility

Resource efficiency attained by closing the material loop has been designated as one of the top priorities by civil society, business, and government in order to reduce resource depletion and related environmental and societal threats (EC 2011; McKinsey 2011; Stahel 1982). On the other hand, taking responsibilities in closing the loop through extended product liability and stewardship of the involved actors has been prioritized by Stahel (1994).

However, the existing scholarly literature on post-retail and closed-loop value chains are limited in certain way. These have yet to explore how the post-retail initiatives and their underlying business components and drivers operate in various fashion businesses characterized by their inherent business philosophies of slow and fast fashion to generate sustainable value. Instead the existing discussion is very generic, for example, on how various fashion companies engage with diverse post-retail initiatives, and so on. However, by investigating whether the trade-off between sustainability and "speed of fashion" (classified into slow and fast fashions) as evident in forward-value chains is equally observable in post-retail initiatives or not would contribute in a number of ways. First, it generates a deeper conceptual underpinning to post-retail initiatives in terms of sustainable value drivers in closed-loop value chains, that is, resource efficiency and liability. Second, by studying and comparing post-retail initiatives of various slow and fast fashion businesses similarities and differences are identified in terms of their underlying business models. Finally, the study opens up an emergent phenomenon underlying sustainable value generation in post-retail initiatives of slow and fast fashion businesses, different from that observed in case of forward-value chains.

2 Fast and Slow Fashion Business Models

The present day fashion apparel industry has increasingly transformed into what is characterized by short product life cycles, high volatility, low predictability, and high impulse purchase decisions by consumers (Christopher et al. 2004; Masson et al. 2007). In such an operating environment, fashion brands and retailers have increasingly started either to embrace the trend and be fast, with short development cycles, and offer large-scale production at a reasonable price, or to counter this trend and promote timeless design, slower rate of production and consumption, and so on (Clark 2008). These contrasting characteristics have resulted in development of two distinct fashion business strategies, models, and philosophies as various scholarly discussions have put it. fast fashion and slow fashion.

Fast fashion is an operational model in retailing predominantly characterized by high seasonal fashion content, affordable prices, frequent deliveries, and large quantities (Caro and Martinez-de-Albeniz 2014; Pacheco-Martins et al. 2014). This results in retailers adopting various strategies such as quick response (Serel 2014), mix of sourcing strategies from nearby and low-wage countries, supplying small quantities of new items at least twice a week (Bruce and Daly 2006), collecting daily sales data and reporting to design departments (Sull and Turconi 2008), constantly adapting design aiming to match the latest fashion trends and customer wishes, and applying flexible procurement policies (Wang et al. 2014). In connection, Bhardwaj and Fairhurst (2010) have characterized fast fashion by the expression of "throwaway," meaning overconsumption and excessive production of short-lived or disposable items. In order to be profitable in such a volatile business environment, fashion apparel retailers need to take the "speed-to-market" approach to capitalize on fashion that is not in the stores of their competitors thus emphasizing responsiveness and agility (Bhardwaj and Fairhurst 2010; Mattila et al. 2002). This has invariably resulted in increasing the pressure throughout the value chain in the recent years, thus increasing criticism from customers, media, and scholars due to the downsides in terms of societal impact (Plank et al. 2012; Pookulangara and Shephard 2014) and environmental problems such as higher carbon dioxide emissions due to frequent shipments and air freight (Choi 2013) among others. Increasingly, these concerns have criticized fast fashion as unsustainable as it encourages disposability, low durability, low quality, and loss of craftsmanship (Cline 2012). Some hard indicators suggest that the clothing consumption level in the western world has increased to almost 15 kg per capita; at this current level of consumption of natural resources we need the regenerative capacity of 1.5 Earths, and by 2050 we will need 2.3 Earths (Deloitte 2013). At the production frontier, the continuous turnover of "cheap" products equates to lower costs overall and higher volume resulting in nearly one-third of the products not sold at full price (Mattila et al. 2002).

In contrast, the ideology of slow fashion challenges the obsession with mass production, instead epitomizing diverse practices such as small-scale production, traditional craft techniques, promotion of local materials and markets, increased use of sustainable, ethically made or recycled fabrics, and development of quality garments with higher longevity (Fletcher 2010; Joy et al. 2012; Wikipedia 2015). In this context, Clark (2008) has provided several examples of local fashion and business practices associated with various aspects of slow fashion.

In many ways this offers a changed set of power relations between the fashion creators and consumers, ensuring long-term relationships and trust through higher degrees of co-innovation and transparency. Fletcher (2010) highlights that a heightened state of consumer awareness of the design process and its impacts on resource flows, workers, communities, and ecosystems can help to bring a systemwide change radically from high-volume, standardized fashion to fewer higher priced exclusive items. In connection with this, various business models have emerged promoting personalization and customization of products—doing it yourself—making, mending, customizing, altering, and upcycling one's own clothing, and so on. This way slow fashion can be seen to promote democratization of fashion not by offering more people access to clothes by lowering prices but by offering people more control over institutions and technologies.

In this context, "fast" and "slow" have become an inherent choice as a proxy for the type of fashion (Fletcher 2010). Whereas "fast" influences ideas of unsustainability, acting as a tool to epitomize increased product throughput and sales thus generating higher unethical and/or environmentally damaging practices, "slow" fashion embraces the ideas and business practices that are logically opposite to "fast," yet administers a systemwide change from the fast-growth model, exemplified by timeless design, high quality and durability, long-term relationships, and better pricing to reflect the "true" cost of the product, thus representing a vision of sustainability in the fashion sector.

3 Sustainable Business Models in Post-retail

Various post-retail activities undertaken by fashion brands, retailers, and small design-oriented brands have emerged in recent times to strive for sustainable value generation (Ekvall et al. 2014). These "re" activities span over five broad areas of reuse, reduce, recycle, redesign, and reimagine, termed 5-R (Ho and Choi 2012). These "re" activities, in various ways, aim at driving higher resource efficiency, lower consumption of virgin materials, and higher end-of-use (EOU) recovery (Niinimäki and Hassi 2011), thus resulting in development of many new business models as one of the key choices for fulfilling a company's sustainable competitive advantage.

A common argument in the literature states that the business model refers to the logic of the company, including how it operates and how it creates value for stakeholders (Magretta 2002; Osterwalder et al. 2005). To date many definitions of business models have become commonly used, but these usually include three key components forming a traditional component-based view. These three business model components are (Yunus et al. 2010):

1. Value proposition: Constituted by the product/service proposed to customers
2. Value constellation: Constituted by the way the company is organized so as to deliver this product and service to its customers
3. Revenue model: Constituting the profit equation meaning how value is captured from the revenues generated

However, this definition of business models predominantly covers the facet of economic value maximization as an outcome of successful business model implementation (in for-profit businesses).

In the post-retail segment the operating business model's value generation spans beyond just generating economic profits, and covers both environmental and social profits. The environmental profit can be considered in line with what is offered in green business models. Bisgaard et al. (2012) highlight the key to develop environmental profits in terms of "substituting to greener inputs, reusing or recycling resources, offering products as a service function while continuing to have ownership of the products, or by developing greener products, services and processes." Tied to this social profits include creation of societal impact of the business in terms of job creation (Stahel 2007), generating social responsibility towards employees, and so on. Consideration of both environmental and social profits in the profit formula results in positioning the business as a change agent in the world but still with sufficient business-like characteristics (Yunus et al. 2010).

4 Strategies for Sustainable Value Generation in Post-retail

Post-retail initiatives in fashion apparel are still new, with limited best practices; the majority of retailers predominantly adopt two main business strategies, either second-hand retailing and/or take-back schemes (Hvass 2014). The literature has linked these strategies to a few generic reverse logistics tasks (Fleischmann and Kuik 2003; Fleischmann et al. 2004):

a. Strategic acquisition (or collection) from consumers
b. Grading (or sorting or disposition) into different fractions on the basis of different qualities and allocation of various reuse options
c. Reprocessing including all transformation processes for future usage
d. Redistribution for delivering the clothes again to the market

Fashion businesses can either run all these reverse logistics operations single-handedly, by arranging both take-back of used clothes followed by reselling them through various retail formats, or can be involved in collaborative networks with a number of other partners to carry out these operations. These result in creating a number of different re-business models for generating sustainable value in used clothing value chains.

Premium and High-Street fashion brands predominantly undertake the strategy of reselling the clothes, either "as is" or after certain redesign, but only with their own branded products. Swedish fashion brands, including Boomerang and Fillipa K resell the used clothes that are deposited by the wearer, either in the same retail stores or in separate stores dedicated to second-hand resale (sometimes in collaboration with other dedicated second-hand retailers). These brands are in this sense closed in their collection of clothes, meaning they only take back their own brand from wearers. Nudie Jeans, another Swedish High-Street fashion brand, further engages with free repair services provided to its wearers in some of its stores (in Stockholm, Berlin, Gothenburg, and London). It also provides a repair toolkit to its wearers that can be ordered online for free. Furthermore, these brands also engage with other operations required to appropriate value of the used clothes, such as washing and redesign. Boomerang, for example, has established a concept called "Boomerang Effect" that includes a return system, a vintage collection (with used clothes meeting certain quality benchmarks and resold "as is"), and redesigning those not qualifying as vintage. These items are totally reconstructed, that is, cut and mixed with production spills to make home interior products as a part of Boomerang's Home section (Niinimäki et al. 2015). Nudie Jeans similarly engages with redesigning its denims: those are beyond repair to make denim rags, by hand-weaving them on manual shuttle looms to attain minimum energy usage.

Alternatively, many fashion retailers—mostly the large, market-driven fast fashion companies—engage with in-store take-back schemes in many ways (Hvass 2014). These take-back schemes could be organized by retailers by entering into donation partnership with charities; for example, Lindex, a Swedish fast fashion retailer, entered into collaboration with Myrorna, a Swedish second-hand retailer run by the Salvation Army (Ekström and Salomonsson 2014). Marks & Spencer together with the charity organization Oxfam have introduced the concept of "shwopping" where wearers who hand in clothes for reuse (in either Marks & Spencer or in an Oxfam store) receive bonus points that can be exchanged for new purchases in Marks & Spencer (Ekvall et al. 2014). The retailers also often give away unsold leftovers or defective clothes to charities. These take-back schemes could also be in collaboration with third-party professional collectors as in the case of H&M and Kappahl, two large Swedish fast fashion retailers. Both have entered into collaboration with Swiss I: Collect (I:Co), a worldwide collector-sorter, which has installed its collection containers in the retailers' store locations. These I:Co containers can usually collect up to 6–7 kg of garments which are then brought back to main storage points by existing logistics, from where I:Co picks them up and transports them to a central sorting plant. Niinimäki et al. (2015) further suggest that take-back schemes can also be organized individually by single companies, as in case of Uniqlo's All-Product Recycling Initiative or Patagonia's Common Thread Initiative.

It can be noted that the above-mentioned business models of second-hand retailing and take-back schemes are associated with multiple R-approaches, such as reuse, recycle, reduce, and so on, and are often intertwined (Niinimäki et al. 2015). Nudie Jeans, for example, engages with repair, reuse, and recycle of its collected jeans which it calls the "Eco-cycle initiative" (Nudie 2015a). Nudie Repair shops help their wearers either by repairing worn-out jeans free of charge or by sending out a free repair kit containing threads, patches, and the like. Even possibilities of breaking down the jeans into something completely different such as a bag, or shorts is possible (Nudie 2015b). However, wearers can also give back the jeans totally, which are then washed and repaired and put back in the shop as a second-hand item. Further in the material loop, if the jeans are totally worn out then Nudie recycles them into cotton pulp or into denim strips to make upholstery (Nudie 2015c). Similarly, charities are also involved with multiple R-approaches, in general selling about 20 % of the sorted items through their own second-hand stores, and the rest are exported for reuse in eastern Europe, Asia, or Africa, or are recycled (Carlsson et al. 2014; Myrorna 2015).

Apparently, these R-approaches together strive to achieve a circular economy where companies aim at closing the material loop by recovering post-consumer clothes and directing them back into the consumption cycle. Jayaraman and Luo (2007) have defined this in terms of four typical reverse logistics loops of reuse, refurbish, remanufacture, and recycle which close the forward material loop by flowing the material back at different points of the value chain. Reuse, for example, redirects the used clothes into the forward-value chain which subsequently demands marketing and sales to coordinate sales of the product "as is". In the case of refurbishing, apart from marketing and sales, outbound logistics is important to bring the product to the service providers, for washing or minor repairing, who subsequently add an appropriate value to ensure that the garment is brought back to a specified quality level and its lifetime is extended. Remanufacture or sometimes called redesign, as an R-approach instead starts with deconstruction of the garment followed by redesigning it into a completely new form. For example, Wardrobe Surgery (now called Re-dress) is a British redesign service provider that completely redesigns used garments on demand from wearers (Re-dress 2015). Many such small initiatives have sprung up with the purpose of redesigning old garments into something completely new. Finally, in Nudie and Boomerang, those products which do not meet a certain quality level are recycled; inbound logistics

Fig. 1 Closing the material
and liability loops

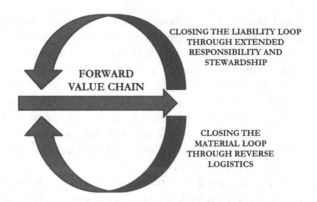

reutilizes different components of the product in conjunction with other raw materials. In this context, it is crucial to note that even though these reverse logistics loops associate different sets of operations they all contribute towards attaining a resource-efficient, circular economy.

Apart from closing the material loop by adopting various strategies to extend the product life and create resource efficiency, Stahel (1994) points out yet another necessity of closing the loop in terms of liability: through extended product responsibility. The next sections discuss these two critical drivers of closed-loop value chains in detail. Whereas driver 1, resource-efficiency, drives for closing the material and product loops, driver 2, extended organizational responsibility (EOR), aims at closing the liability loop by extending the stewardship of the involved actors in the used clothing network, as shown in Fig. 1.

5 Driving Factor 1: Resource-Efficient Reverse Logistics for Closing the Material Loop

Resource efficiency in the process of closing the product and material loops highlights the attempts taken by actors for subsequent dematerialization (Stahel 1994; Tukker 2015) by:

1. Prolonging the service lifetime of products
2. Ensuring products are used as intensively as possible
3. Manufacturing products as cost and material efficiently as possible
4. Reusing products as far as possible after the end of the product's life

Stahel (1994) highlights that these activities could lead to a minimization of material flows in the economy while maximizing service output or user satisfaction, thus generating a self-replenishing system. In such a system the resource input per unit use over long periods of time can be optimized thus leading to a structural change in the economy. Apparently Stahel's work shows that the smaller the loop (as in the case of reuse) the higher is the profitability. The inherent drive for a resource-efficient society and businesses come from the inevitable consumption growth rate and the competition for resources in a resource-constrained world. The European Union (EU) has therefore designated resource-efficiency as one of the flagships of its Europe 2020 strategy (EC 2011). In this context, it is crucial to note that clothing is a material-intensive product and depends upon huge quantities of natural resources required for production, distribution, and maintenance, coupled with the consumer's current desire for excessive consumption (Armstrong et al. 2015). A study by Farrant et al. (2010) on the environmental benefits of reused clothes showed that consumption of natural resources such as natural gas and crude oil both can decrease by about 15 % if clothes are reused through second-hand shops (in Estonia and Sweden) instead of being directly disposed of after first-hand usage. The study further concluded that the environmental burden of the product life cycle is reduced by almost 14 % (in terms of global warming) through reuse of clothes.

Several commercial business strategies and resource-miser business models have been emphasized in Stahel's works (1994), (2007) including redesigning, reusing, rebuilding, technology upgrading, and the like to reduce the volume of material through the economy. Fashion businesses, however, have adopted a resource-efficiency strategy in a quite complex way. The fashion retailers and brands have adopted various strategies (as mentioned above) to attain higher resource efficiency along the R-approaches, however, at the same time provoked wearers to purchase more by providing discount vouchers on return of used clothes. Ekvall et al. (2014) have referred to this aspect in terms of displacement rate, meaning the potential of a used item to replace the purchase of a new one. The displacement rate is critical for evaluating the magnitude of environmental gains offered by various reused business models. Typically the potential for displacement, hence the degree of environmental gain, is influenced by various factors, such as quality of the resold items, percentage increase in "usage time," and discounts offered on new purchases.

Fashion brands such as Nudie Jeans and Fillipa K are engaged with reselling of their own brand and could be claimed to have a higher displacement rate due to the higher quality of resold items and higher product price compared to the average resold products. However, a WRAP study from Britain (WRAP 2013a) has indicated that the reuse displacement rate from buying a used item rather than a new one is only 28 %, meaning three reused items can offset the purchase and thereby the production of just one new item. Ekvall et al. (2014), however, highlight that this result is for average quality second-hand clothing and could possibly be higher for luxury second-hand compared to second-hand clothes in charity shops.

It is evident that the scholarly literature in exploring the exact impacts of displacement is quite shallow. An alternative viewpoint that exists is that this direct displacement of new purchases does not always occur in reality (Ekvall et al. 2014). Ekvall and his colleagues note in a recently published Norden report called "EPR Systems and New Business Models" that even if high-quality used clothes can enhance the functioning lifetime of the item (called the usage time), in the meantime a user may have saved enough money to purchase a new shirt that causes environmental impacts thus reducing the environmental gain. Ekvall et al. (2014) states that:

> Usage time is taken to mean the intensity of use multiplied by the period over which it is used. An article of clothing can last many years in someone's wardrobe without being used. In this case the lifetime might be high but the total usage time would be low. It is the usage time of an article of clothing which is key in how much it offsets the purchase of new similar clothing articles.

Furthermore, fashion retailers and brands including H&M, Kappahl, Boomerang, and many others have started offering discount vouchers to wearers on return of clothing through the take-back schemes. These discount vouchers typically offer 10–15 % off on the next purchase of new items and do not necessarily cause displacement, and in fact may lead to higher desire for a new purchase. Simply speaking, such offers may create a perpetual voracious cycle of buying new clothes using discount vouchers obtained by depositing old ones thus adding to the throwaway culture even more. These issues have eventually made the effects of displacement rather complex to underpin, and hence can only be speculated.

6 Driving Factor 2: Extended Organizational Responsibility for Closing the Liability Loop

Extended producer responsibility (EPR) is defined based on two principles (OECD 2001):

1. Shifting of responsibility (physical and/or economical; full or partial) upstream toward the producer and away from municipalities
2. Provision of incentives to producers to take into account environmental considerations while product designing

EPR originated in the 1990s as an environmental policy strategy through an analysis of a number of Swedish and foreign recycling and waste management programs, as well as the use of various policy instruments to promote cleaner production (Lindhqvist 1992; Lindhqvist and Lidgren 1990). The intention behind the concept was to create a framework for governmental legislation or regulation based on the "polluter pays" principle (Ekvall et al. 2014). By definition both upstream and downstream effects are included in EPR schemes which aim for producers to gain specialized expertise (e.g., product design, materials, or technology development), utilize new resources (financial and technical), and stimulate and educate their customers to accept alternatives to landfilling and incineration and to participate in waste product recovery (Lindhqvist and Lifset 1997).

However, trade-offs exist in driving these schemes in terms of being either mandatory or voluntary, upstream or downstream, and so on in determining the scope of participation of various actors in the clothing network (Ekvall et al. 2014; Kibert

2003). Furthermore, shifting the responsibility towards the use phase of the product is not exclusively linked to original manufacturers or producers but may also include other actors, such as charity organizations and service providers, within the product's value chain. Initiatives undertaken by brands, retailers, and other actors in the clothing network, involving product take-back, reselling, repairing, upgrading, and the like commonly extending product stewardship (Kostecki 1998), fall under the broader scope of extended organizational responsibility (EOR). Such industry-driven EOR where various actors of the value chain are proactively involved, focuses on capturing untapped value from used clothing (Hvass 2014).

According to Lindhqvist (2000) and Tojo (2004), extended responsibility of an EPR scheme can be categorized and evaluated along five different perspectives. These are:

i. Liability
ii. Economic or financial responsibility
iii. Physical responsibility
iv. Informative responsibility
v. Ownership

Liability includes the responsibility for detectable environmental damages related to a specific product. The degree of liability is, however, limited by legislation and therefore depends on the different national and regional laws and does not lie directly in the hands of the value chain actors. Presently, only a few policy-driven liability schemes exist in clothing, such as France's eco-TLC and Canadian legislation expected to commence in 2017 onwards (Kelly 2012). Next, economic or financial responsibility is used to describe an actor who will fully or partly cover the costs of collecting, recycling, or final disposal of the products and these costs could be paid either directly or through a special fee. On the other hand, physical responsibility refers to an actor taking part in the physical management of the products or the impacts of the products (Lindhqvist 2000; Lindhqvist and Lidgren 1990). Lindhqvist (2000) further defines informative responsibility as the multiple possibilities for an actor to supply information on the environmental properties and effects of its products and create awareness and understanding. Finally, if the actor retains the ownership of its products over the entire life cycle then it is automatically coupled to the environmental impacts of the products. These responsibilities are, however, not exclusive to producers alone and are equally relevant for other actors in the value chain, forming the broader applicability of EOR.

7 Empirical Study

The present study employs a qualitative research conducted through in-depth, semi-structured interviews and document studies to build suitable cases for the purpose of an explorative inquiry (Cresswell 2007). Extant discussion on each of the constituent concepts used in this study, viz. extended responsibilities for closing the liability loop, resource efficiency for driving a circular economy, along with the generic discourse on slow and fast fashions in terms of sustainability are quite established, hence a deductive approach is followed to underpin these. However, the discourse on sustainable value generation by slow and fast fashion businesses in the post-retail segment has not yet been explored thoroughly, thus advocating a theory-building perspective (Eisenhardt and Graebner 2007).

The study was conducted by collecting data from 12 fashion companies each operating in the post-retail segment in various ways. The case companies were chosen through purposeful sampling to represent the dominant fashion business models operating in the post-retail segment (Pal 2015). These are namely, fast fashion retailers, slow and fast fashion brands, and smaller redesign brands. Table 1 describes the cases in detail in terms of their post-retail initiative and underlying business model components.

Primary data were gathered through a single semi-structured interview conducted with each case company. All the respondents held top decision-making positions related to post-retail initiatives in their respective companies (e.g., sustainability manager, operations manager, or owner/managing director). Such strategic roles and responsibilities of the respondents justified their viable reflection on the topic of the present research, with low intracase differences in opinion. Each interview lasted between 45 and 60 min and was conducted in English. Each of the interviews was later transcribed and coded along themes used in both Table 1 (post-retail initiatives and underlying business model components) and in Table 2 (displacement effect and EOR).

The interview questions aimed at gaining rich descriptions of the current post-retail initiatives along with detailed insights on the responsibilities undertaken by each case company. The respondents were asked to describe their company's key post-retail activities, what they offer to the customers, and how revenue was generated. In addition, questions covering EOR topics, such as ownership, type of sustainability communication, and information sharing in post-retail initiatives, collaborations with other actors, and so on were also asked.

Additionally, written documents and reports acquired through search engines were also analyzed in order to both support and validate some data obtained through the interviews. In particular, over 700 pages of reports published by the Nordic Council of Ministers, called the Norden reports, were scrutinized to supplement the data gathered. These reports are available online and free through the Norden publication database.[1]

For data analysis, the transcribed interviews were deconstructed to generate relevant cues; those could be categorized under the two main themes constituting sustainable value generation in post-retail initiatives, that is, extended responsibility and displacement effect. Some minor modifications have been introduced in the concepts underpinning this study to refine the deductive framework and suit it to the research context and purpose.

---

[1]Norden Publication Database: http://norden.diva-portal.org/smash/search.jsf?dswid=-1454 (02.09.2015); an extended list of these reports can be available from the author on request.

Table 1 Case description with business model explanation

Cases	Post-retail initiative	Business model components
Monki/H&M Swedish fast fashion brand retailer	In collaboration with a global collection, sorting and recycling firm, I:Collect (in short I:Co), Monki engages in taking back of used clothes under its "Second Chance" program through its 22 stores. It offers strategic locations in its shops for setting up I:Co's collection containers. In return gives discount vouchers to customers	*Key involved activities* • Strategic collection and setting of partner's collection containers in shops *Customer value proposition* • Discount vouchers up to 10 % to old wearer on return of used clothes (applies to purchases over 300 SEK[a]) • Customers can receive a variety of rewards • Sustainable image of extended responsibility *Profit/revenue generation formula* • Fresh purchase of new clothes by customers using discount vouchers • Money from I:Collect
KappAhl Large Swedish fast fashion retailer	In collaboration with a global collection, sorting and recycling firm I:Co, KappAhl engages in taking back of used clothes under its "Wear, Love, and Give Back" program. It offers strategic locations in its shops for setting up I:Co's collection containers. In return gives discount vouchers to customers	*Key involved activities* • Strategic collection and setting of I:Co's collection containers in shops *Customer value proposition* • Discount vouchers of 50 SEK per donated bag to old wearer on return of used clothes (applies to purchases over 300 SEK) • Sustainable image of extended responsibility *Profit/revenue generation formula* • Fresh purchase of new clothes by customers using discount vouchers • Money from I:Collect
Nudie Jeans Swedish fashion brand	• Engages with repair of its own brand (wearer retains the ownership) • Takes back own brand from old wearers for reselling, repurposing, or recycling and in return gives discount vouchers, under its "Eco-cycle" program consisting of Repair, Reuse, Reduce	*Key involved activities* • Repair services • Reselling in own stores; also includes picking and packing, laundering, and repairing • Clothes not qualifying for resale are either used to make denim rags or sent for recycling *Customer value proposition* • Extended active lifetime of the clothes, minor redesigns • Free repair service in Nudie repair shops or free sewing kit • Resold branded clothes at lower price • Reducing consumption of new clothes *Profit/revenue generation formula* • Purchase of second-hand clothes and denim rags

(continued)

Table 1 (continued)

Cases	Post-retail initiative	Business model components
Boomerang Swedish fashion brand for casual wear	• Takes back own brand from old wearers for minor redesigning or repurposing, and reselling through own stores, under the concept called Boomerang effect launched in 2008. Old wearer is offered discount voucher on return	*Key involved activities* • Reselling in some of its regular stores (with new clothes) as Boomerang Vintage • Redesigning into interior products (Boomerang Home) for selling as home textiles in own store *Customer value proposition* • Resold branded clothes at lower price • Partly reducing consumption of new clothes • Discount vouchers of 10–20 % to old wearer on return of used clothes; old wearer should include the personal life story of the clothing *Profit/revenue generation formula* • Sale of second-hand clothes as vintage and of interior products
Haglöfs Swedish outdoor clothing brand	• Takes back own brand from old wearers for reselling, under the concept of "Swapstories" and in return gives discount vouchers to the old wearer	*Key involved activities* • Reselling in Haglöf's brand store (with new clothes) under the second-hand concept called "Swapstories" as long as the clothes are intact and clean *Customer value proposition* • Discount vouchers of 20 % to old wearer on return of used clothes • Partly reducing consumption of new clothes *Profit/revenue generation formula* • Sale of second-hand clothes
Beibamboo Finnish baby clothes brand with leasing option	• Involved with leasing of clothes (presently nonoperating in this format); retains product ownership	*Key involved activities* • Leasing of clothes • Delivery and pick-up services • Laundering (as an additional service) *Customer value proposition* • Collaborative usage of baby clothes • High product quality to ensure extended lifetime *Profit/revenue generation formula* • Rental fee

(continued)

Table 1 (continued)

Cases	Post-retail initiative	Business model components
Uniforms for the Dedicated Swedish slow fashion brand creating timeless design using recycled fibers	• Involved with designing and producing environment-friendly fashion garments based on recycled fibers • Rents a small part of its collection on a short-term basis, under the concept "The Collection Library" • The Rag_Bag concept for donating old clothes	*Key involved activities* • The Rag_Bag concept • Leasing of clothes *Customer value proposition* • Collaborative usage of leased clothes • Sustainability image *Profit/revenue generation formula* • Prepaid ragbags • Rental fee
Dream and Awake Swedish redesign brand and studio	• Sells unique redesigned clothes • Organizes redesign workshops with individual wearers (wearers retain the ownership of the product)	*Key involved activities* • Old clothes are collected, redesigned, photographed, and sold through mobile studios and online shops • Redesign services through mobile studios (design, facility, equipment) to wearers *Customer value proposition* • Upcycling through redesign • (Re)-design co-created with wearer *Profit/revenue generation formula* • Sale of redesigned clothes • Redesign charges
Stormie Poodle Swedish designer-based children wear brand	• Upcycles of high quality hotel linen into kids wear	*Key involved activities* • Buying of washed linen from laundry, initial sorting, product development, organizing sorting and manufacturing at vocational facilities *Customer value proposition* • Value creation from industrial wastes *Profit/revenue generation formula* • Sales of repurposed products through web shops and also to retailers

(continued)

Table 1 (continued)

Cases	Post-retail initiative	Business model components
Design Stories Swedish designer-based home interior brand	• Upcycles clothing wastes and production spills of project partners (such as charities, second-hand retailers, etc.) • Organizes do-it-yourself (DIY) workshops	*Key involved activities* • Collection, design, and prototype development from clothing wastes and production spills of project collaborators; outsourced production • Co-developing repurposing processes through communication via various workshops, tutorials and seminars *Customer value proposition* • Value creation from both production and consumer waste materials *Profit/revenue generation formula* • Online sales of repurposed products • Workshops held at companies
Mocklis Swedish manufacturer of folklore and knitted socks (with timeless design)	• Engaged with upcycling of leftovers from Swedish charity organization	*Key involved activities* • Picking of usable materials from leftovers, cutting, and repurposing (outsourced sorting, laundering, and sewing processes) *Customer value proposition* • Value creation from consumer waste materials • Social sustainability in engaging prisoners in workforce *Profit/revenue generation formula* • Online sales of repurposed bags
Skryta Swedish designer-based home interior brand	• Offers design services to upcycle both post-consumer clothes waste obtained from Swedish charity organization and production spills	*Key involved activities* • Picking of usable materials from leftovers, cutting and repurposing (outsourced sorting, laundering, and sewing processes) *Customer value proposition* • Value creation from both production and consumer waste materials • Social sustainability in engaging prisoners in workforce *Profit/revenue generation formula* • Sales through small web shop for local designers

[a]*SEK* Swedish Krona

Table 2 Qualitative rating of closing the material and liability loops by various business models

Business model category	Displacement potential[a]	Extended organizational responsibility[b]
Fast Fashion brands and retailers (Evident cases: Monki, KappAhl)	Low	Medium (1.5) [IR = Partial (0.5), PR + OFR = Full (1), O = No (0)]
Slow Fashion brands (Evident cases: Nudie Jeans, Boomerang, Haglöfs)	Low or Medium	Medium-to-High (2) [IR = Full (1), PR + OFR = Full (1), O = No (0)]
Slow Fashion brands with leasing (Evident cases: Beibamboo, Uniforms for the Dedicated)	Medium	High (3) [IR = Full (1), PR + OFR = Full (1), O = Full (1)]
Slow Redesign brands (Evident cases: Dream and Awake, Stormie Poodle, Designstories, Mocklis, Skyrta)	High	Medium (1.5) [IR = Partial (0.5), PR + OFR = Full (1), O = No (0)]

[a]Displacement effect is denoted as Low/Medium/High depending upon whether the businesses do/partially do/do not offer a scope to purchase new clothes through its business
[b]Extended organizational responsibility is denoted as Low/Medium/High depending upon whether the businesses takes no/partial/full liability in managing the post-retail initiative. It is cumulative regarding three constituent responsibilities: information responsibility, physical + outsourced financial responsibility, and product ownership

First, resource efficiency forms a very crucial base in this study for defining sustainable value generation in post-retail businesses, and this is measured in terms of displacement potential. The displacement potential is recorded along a three-point Likert scale (high–medium–low) depending upon whether the post-retail initiative does or does not offer any possibility of purchasing new clothes through its business. The displacement effect is high if no such possibility is offered, for example, in case of reuse, and it is low if discount vouchers are offered to wearers on return of old clothes. However, the correlation is much more complex in reality as there are other intervening factors deciding the displacement potential.

Within the scope of EOR, considering that there is no current legislation for producers or importers to take responsibility for the environmental impact, waste management, and disposal of their products (in the countries from which the cases have been selected), liability as a responsibility was not further investigated. Additionally the financial responsibility was adapted into outsourced financial responsibility in order to differentiate those actors who solely pay third parties to take care of their products. Even though physical and informative responsibilities can be strictly executed by one actor in the value chain itself, they are always connected to labor and resource costs, and therefore finances, which are not considered in the existing literature. Thus based on this logic one could argue that every actor that takes over physical and/or informative responsibilities is also financially responsible. The extended responsibility was measured along its three components, information responsibility (IR), physical and outsourced financial responsibility (PR and OFR), and ownership (O) using a similar three-point Likert scale (high–medium–low). If cues from the transcribed interviews or written documentation were obtained supporting each of these components then they were scored as 1, otherwise 0. This denotes whether the company takes no/partial/full liability in managing the post-retail initiative.

8 Post-retail Initiatives and Underlying Business Models

8.1 Key Activities in Post-retail Initiatives

The study showed that the fast fashion businesses (Monki/H&M and KappAhl) predominantly engage with product take-back schemes through strategic collection of used clothes in their stores. Such activities are organized in collaboration with global sorting partners such as I:Co or Kicki. Collection takes place in containers of 6–7 kilo capacity and on deposition wearer receives a discount voucher for use during a subsequent new purchase.

The slow fashion brands, on the other hand, engage with a wider range of activities, including collection, refurbishing through repairing, laundering, and so on, and reselling. These brands generally organize these multiple reverse logistics activities under a marketable concept or program; for example, Nudie Jeans under its "Eco-cycle" program offers its wearer scope to repair broken jeans either at the Nudie Repair Stores or by using DIY repair kit, free of charge. Through its website Nudie offers its wearers the possibility to fill in a form to dispatch the repair kit needed to mend the broken jeans. Such a repair kit typically contains denim patches, iron patch, needle, spool of thread, repair kit booklet, and thimble. In addition, reselling is also organized by these slow fashion brands either in the regular stores, as in the case of Boomerang and Nudie, or in different stores dedicated to selling second-hand items of their own brand, as done by Filippa K. Furthermore, these brands also take care of washing the garments, and picking and packing them before they are put out for sale. Boomerang organizes such resale through seven stores selling its own brand, four of which are located in Stockholm. Ekvall et al. (2014) report that since the start in 2011 Boomerang has collected around 7000 garments annually. Nudie Jeans similarly gives a reuse option to its wearer, by offering possibilities either to make minor patchworks to extend the life of the jeans, or to make something completely different by reconstructing the jeans into a bag or shorts. Otherwise, wearers could donate their old jeans which are then washed and repaired and put back in the shop as second-hand items only if they achieve the Swedish "Good Environmental Choice" eco-label standard. Sometimes, Nudie Jeans is also involved in collaborations with designers and other creatives under the program "Denim Maniacs" to give worn-out jeans a second life. Haglöfs engages with a similar program aimed at reselling their own brand as second-hand. Some of the other slower fashion brands have ventured into leasing and renting business models, in addition. For example, two such brands observed this study—Beibamboo and Uniforms for the Dedicated—engage with leasing a specific section of its clothes. Alongside they are involved with other service provisions necessary to appropriate the leasing activity, such as laundering, delivery, and picking services. However, due to several management reasons Beibamboo discontinued its leasing concept 3 years after its start. During its operation period, the business had about 25 customers who could either opt to rent a mini- set with 6 pieces of baby clothes or a basic set with 15 pieces (Ekvall et al. 2014). Uniforms for the Dedicated, on the other hand, is involved with renting a small part of its collection consisting of selected suits and outerwear on a short-term basis (for a maximum of a week) through its store in Stockholm.

The smaller slow fashion, on the other hand, engages more with redesign activities. Although some of them are totally based upon the redesign philosophy, as in case of Dream and Awake and Stormie Poodle, the others include some redesign alternatives in their regular slow design collections. The redesign activities are mostly conducted with used clothes or textiles, either bought or collected from other actors in the used clothing network or individual wearers. Stormie Poodle, for example, works with reconstructing hotel linen into kidswear by buying washed linen from the laundries followed by initial sorting of the material in Sweden. The linen is then sent to Latvia for a second sorting in collaboration with vocational schools. Dream and Awake, on the other hand, collects or buys old vintage clothes from the market and redesigns, photographs, and finally sells them through mobile studios. It is also involved with organizing redesign workshops with wearers in providing designs, facilities, and equipment to help them redesign their own clothes. Working with a slightly different format, the slow fashion brands organize redesign activities by working with charities including the Red Cross, to collect textile wastes and leftovers from them. These materials are then redesigned to develop prototypes in the studios of these redesign brands. Other supporting activities such as sorting, washing, and sewing are usually outsourced to other organizations including social institutions such as prisons and disability homes. Mocklis, for example, works in collaboration with Syverket, a company that sorts the wastes thus helping the brand to pick up the usable materials; this is followed by organizing sewing activities at the workshops in prisons. In addition Mocklis also utilizes small local production facilities to produce the final redesigned products (bags). Design Stories, in a similar way, works with another charity-led organization called Emmaus, to get diverse materials such as plastic bags and fabric. It collects production spills from other companies to fuse these materials to make slow craft hand-made lamps.

8.2 Customer Value Proposition (CVP) of Post-retail Initiatives

The fast fashion businesses aim at generating a sustainable brand image to their customers through involvement with post-retail initiatives. Both Monki/H&M and KappAhl share extensive information about their collaborative garment collection schemes through company websites and social media to show their commitment towards making a change in the garment life cycle. Through their programs, "Second Chance" and "Wear, Love, and Give Back," respectively, these businesses quite explicitly share information about the post-retail objectives as an integral part of company values. Monki, for example, clearly specifies "Second Chance" in their website under company values and as an integral part of their "way of doing business." KappAhl (and also H&M), on the other hand, considers and reports such initiatives under "Our Responsibility" in their websites. In addition, fast fashion businesses also offer discount vouchers to all wearers on donation of old clothes under different schemes, either as 10–15 % off or as a monetary discount (50 SEK) on new purchases over 300 SEK.

The slower fashion brands, such as Boomerang, Nudie Jeans, and Haglöfs, also engage in a similar way to offer value to their customers, by offering discount vouchers ranging between 10–20 % to the old wearer on their return of used clothes. In addition these brands also try to tell a story to its wearers through the process of reselling. Haglöfs, for example, under its second-hand concept of "Swapstories" narrates a story written by the garment's previous owner about what they have experienced together with the product (Reuters 2012). Such stories can be submitted in advance via the Swedish Haglöfs website, where it is possible to read other people's stories as well. On submitting a story along with returning a garment a higher discount can be obtained. Nudie Jeans, on the other hand, maintains one of the most comprehensive websites where it shares detailed information on every step of its "Eco-cycle": break-in, repair, reuse, and recycle. Nudie's storytelling to its wearer is based upon a transparent inscription of all its processes and operations through blogs, videos, and guides, and flowcharts posted through its website, thus offering a unique value and sense of responsibility to its wearer. Free services in the form of in-store repairing and sending out of repair toolkits are a part of the extended value proposition and brand strategy of Nudie Jeans. As highlighted by Niinimäki et al. (2015) both Nudie Jeans and Boomerang use this strategy to endorse their brand and communicate sustainability. To some extent such transparency in communicating information explicitly portrays these brands' initiative to attain a circular economy and resource efficiency.

On the other hand, the fashion brands engaged with leasing activities focus on offering their wearers the benefits related to collaborative consumption. In case of Beibamboo, typically the benefit offered to the babies' parents was in terms of payment of a small fee to receive a set of 5–12 clothes rented over a long period of time, till the babies outgrow the size. Additional services such as delivery and pick-ups were also arranged by the brand along with washing and regular care of the clothes.

In general the redesign brands, however, focus more on the redesign aspects of the product through upcycling. These brands claim to reinvigorate and redefine the old into something new. Dream and Awake, for example, claims that in doing so it adopts a "social + green" business model format as it uses no virgin material during its production and the social aspect is maintained by paying a fair sum of money to the tailors. Along with this, the wearers are also offered the feeling of wearing "unique pieces," available in one sample only, very similar to having a customized vintage garment. Stormie Poodle similarly generates a combination of the "social + green" business model by utilizing old hotel linen as the starting raw material and further carrying out the production at vocational facilities in the Baltic countries to support social development. Mocklis and Skyrta, on the other hand, build upon their social aspects of sustainability by engaging prisoners from the jails and handicapped people. In all, these brands highlight their resource efficiency and circularity through upcycled design to underpin their sustainability image, thus proposing value to the wearer or user.

8.3 Revenue Generation from Post-retail Initiatives

Revenue generation is a critical requirement of all business models. Such revenue generation in "social + green" business models stands on three pillars: social, environmental, and economic profits. In the case of the post-retail initiatives, balancing these three profit formulas significantly keeps the "social + green" business model operating sustainably. Although this section only highlights economic profit as a component of traditional business models, the environmental and social profits are reported as sustainable value generators in closed-loop value chains, in the next section.

The fast fashion companies receive some money from the sorting partners in supporting the in-store collection activities although most of it is donated to charities and for societal causes. Monki, for example, reports that it receives money from I:Co even though any earnings made are split between H&M's research fund and Plan International's "Because I Am a Girl" project. However, a big share of revenue is generated indirectly from the post-retail business format by offering discount vouchers. Such discount vouchers are expected to generate higher purchases of new items thus resulting in sufficient revenue generation.

The slow fashion brands mainly generate revenue in two ways. Similar to the fast fashion companies these brands also provide discount vouchers to the wearers thus creating an indirect revenue generation stream from their post-retail initiatives. Apart from that, resale from second-hand clothes sufficiently generates revenue as well. Boomerang, for example, sells its second-hand vintage collection through regular stores at a price range between 300–500 SEK, whereas the Boomerang Home products in the Effect Collection made from fabric offcuts are sold at a much higher price. Similarly, Nudie Jeans sells its rag rugs made from denim rags at a price range between 1999–5999 SEK. However as reported in Palm et al. (2014), Haglöfs did not gain any money from the sale of its second-hand items due to its high amount of charitable donations (Palm et al. 2014).

The redesign brands in a similar way price their upcycled products quite high as revealed through the study. Mocklis, for example, priced its upcycled bags between 1000 to 2200 SEK whereas Skyrta sells its lamps on-demand (made out of Red Cross's leftovers) at over 5000 SEK per piece. However, most of these small-scale redesign brands work on the redesign business as either an experimental or start-up project alongside their main business model of selling slow fashion design, mainly due to the long break-even periods in the upcycling business and other inherent challenges to it, such as high costs of sorting and low recovery rate among others. Stormie Poodle, for example, highlights the high cost of redesigning in Sweden and the long manufacturing time required to redesign each product manually.

9 Post-retail Initiatives and Closing the Loop

9.1 Closing the Material Loop by Displacing New Purchase

Fast fashion businesses communicate their post-retail collection initiatives in collaboration with other partners as an attempt to close the material loop to attain a resource-efficient circular economy. However, several recent investigations criticize these efforts, considering them to be merely a part of branding and marketing (Ekvall et al. 2014). The recent debate on whether fast fashion companies are actually sidestepping the issues of overproduction and consumption by showcasing these post-retail initiatives is steadily gaining prominence (Guardian 2015). In this context, the potential to displace the purchase of new garments serves as one of the possible indicators of such a closed loop process. However, this is one of the major drawbacks of the attempt to close the loop undertaken by the fast fashion companies. They engage just with collection of the used clothes and in return provide discount vouchers worth 10–15 % off from the new purchase from their own stores, resulting in a low degree of displacement potential. Customers are now simply attracted towards a perpetual environment of discounts available year-round in the stores: one of the key drivers of a throwaway economy thus depositing more and more of their old clothes and in return buying new ones. With limitations of the currently available technologies complete separation of the mixed fibers into constituents is still a tale of the future, and thus does not really support the claim that recycling the donated clothes would rechannel the entire material into the resource pool once again at the same rate at which virgin materials are utilized. Moreover, each turn of fashion also creates a huge carbon footprint, even if it's in the loop.

The slow fashion brands in spite of offering similar discount vouchers to the customers on donation of their used clothes further engage with resale of second-hand items. Such resales not only extend the lifetime of the old garments but are expected to replace the purchase of new items as well thus having considerable potential to displace new purchases. Furthermore, such reuse material loops are expected to have considerably lower carbon footprints and impact on natural resources (Farrant et al. 2010). The leasing slow fashion brands go a step farther in their effort to increase displacement of new purchases. They possess the positive potential as also rendered by resale through extension of the usage lifetime of the product. In addition, they promote collaborative consumption thus satisfying multiple users, as also highlighted by Mont et al. (2006). This significantly increases the usage time of each product (as a ratio to its shelf life) thus having the potential to have a higher displacement rate. However, this relationship is not simple as in the meantime the wearer/shopper may save enough money to buy a new one, thus upsetting the benefits that displacement gained.

In comparison, the redesign brands offer a much higher displacement effect as they rarely utilize virgin materials to manufacture new items. Instead these brands are constantly upcycling old used materials and rechanneling them into the consumption cycle, thus having significant impact.

Overall, in terms of resource efficiency (measured by the potential to displace the purchase of a new one), the redesign brands show the best results in terms of displacement potential: to replace the purchase of new clothes by the resale of redesigned old clothes. The slow fashion brands are also considerable in terms of their displacement potential countering the negative impacts of offering discount vouchers to stimulate new purchases. However, the fast fashion companies prove to be the worst among all considering their sole endpoint to closing the loop activity is by stimulating new purchases.

9.2 Closing the Liability Loop Through Extended Organizational Responsibility (EOR)

The fast fashion companies, Monki and KappAhl, show low physical responsibility in managing used clothes beyond their in-store collection in the I:Co containers. Responsibility is limited in terms of just offering space for placement of the collection containers. However, further stewardship of sorting and reselling/reusing/recycling these collected items is financially outsourced to I:Co through a strong partnership. I:Co maintains a well-established automatic collection and logistics network to transport the collected used clothes to its Wolfen Textile Recycling Plant. It is still unclear "who pays whom" in this collaboration. Interviews suggest that it is the brands and retailers who pay a fee to I:Co to install these containers thus undertaking a financial responsibility in outsourcing the physical management of the waste. On the other hand, company websites suggest otherwise; for example, Monki reports that it receives money from I:Co even though any earnings made are split between research and charitable work.

Information responsibility in the case of these fast fashion companies is communicated very clearly through company websites, marketing campaigns, and other social media. This has become an integral part of their brand strategies and an avenue to communicate their sustainability image. Most of the time, the information on post-retail initiatives and collection schemes are documented either in the company's sustainability report and can be found under "core value" in the company's website. Monki, for example, conveys information about its "Second Chance" program, along with other corporate social responsibilities under "Monki

Values". However, interestingly, none of these fast fashion companies provides a transparent and holistic view of the entire closing the loop process, meaning that it is unclear what happens to the collected items once they are received by I:Co. Do they end up in Africa or some other developing regions, and so on? The figures projected are also very coarse; for example, ~30 % is recycled as stated on I:Co website. The fast fashion companies do not refer to details of traceability of the products once collected through their stores, even though they claim to disagree on this issue. Some fast fashion companies do deposit their leftovers to I:Co which can be traced back from their brand logos, until they are debranded.

The slow fashion brands, Nudie Jeans, Boomerang, and Haglöfs, on the other hand, exercise a high degree of physical responsibility towards managing the used clothes of their own brand collected through their stores. Such liabilities as discussed earlier include collection, refurbishing services, along with reselling and recycling thus representing multiple reverse logistics processes. These brands sell the clothes as second-hand directly through their own stores only after they attain a certain quality standard. Those that do not pass the quality test are instead converted to a completely different product category. Nudie Jeans, for example, engages with collection of its own used brand through its stores, provides free repair services to wearers in case they want to extend their period of use of the jeans, and in other cases takes back the jeans from the wearer to wash, repair, and put them up for resale. Only those jeans which have achieved the Swedish "Good Environmental Choice" eco-label are resold, and the rest are recycled to make something completely different out of the denim rags. Nudie's recycled product range includes carpets and camper seats. Similarly Boomerang's concept of "Boomerang Effect" includes a holistic management of their used own brand through a return system, reselling vintage and redesigning the rest into interior home decors.

Communication of information related to such comprehensive post-retail initiatives is also done in a very systematic way. Niinimäki et al. (2015) suggest that wearers and shoppers who are actively interested in sustainability look for related information and news primarily through the company websites apart from in-store communication of sustainability. Lack of special campaigns and immediate first-page information on the company website invariably results in not capturing the attention of the less proactive customers. In this context, Nudie Jeans is a pioneer in communicating its efforts transparently through its website. It has a very dedicated website that conveys all its post-retail initiatives and processes. Not only do all these brands use their regular stores to communicate their story to the wearers, but Nudie by setting up its repair shop inside the regular shop, and Boomerang and Haglöfs by mixing reselling of second-hand with new ones in regular stores also reach out to the consumer. In addition, Nudie Jeans also offers its wearer a multitude of information, both audiovisual and readable, including washing guides, self-repairing booklet, end-of-life possibilities, and the like to communicate the services and other actions for becoming more sustainable. Overall, these brands engage in a very high degree of information responsibility.

In an alternative business model based upon leasing, Beibamboo exercises complete physical management of the clothes to and from each wearer along with its maintenance through washing and minor repairs if required. Furthermore, under the leasing model the brand also retains complete ownership of the product throughout its lifetime. Similarly in autumn 2013 Uniforms for the Dedicated also selected some parts of its collection for short-term leasing in its store in Stockholm thus contributing towards communicating careful shopping (Niinimäki et al. 2015). It also communicates such efforts through its official website, and the rental concept is labeled as "The Collection Library." Here interested renters can read about the idea of what it calls "time share" and can get all practical details and terms related to renting a Uniforms for the Dedicated product. In addition, its owner Mike Lind considers the use of store as the platform to communicate these efforts to customers as not many of them are totally aware of the concept.

Redesign brands are also at the forefront of taking physical responsibility of the wastes and used clothes that are neither produced nor imported by them. However, this is executed in a slightly different way. These companies either buy or receive various kind of wastes, such as clothing wastes (in the case of Mocklis, Skryta, and Dream and Awake), production spills (in the case of Design Stories), and used textiles (in the case of Stormie Poodle) from a number of suppliers or partners including charities, retailers, or even individual people. These suppliers or partners can be commercial, such as the hotels supplying used linen in the case of Stormie Poodle or nonprofit charities in the case of Design Stories, Mocklis, and Skryta. After collection, these small redesign brands engage with design and prototype development. Considering the relatively small size of business, these redesign brands have limited financial and infrastructural leverage, hence they are compelled to outsource many of the value-appropriating services, such as sorting, washing, and sewing required to regenerate value of the upcycled product to prisons or to other service-providing companies.

Furthermore, information responsibility of these small-scale redesign brands are predominantly through physical platforms, such as training sessions and workshops as compared to communication through websites. Dream and Awake, for example, communicates its redesign potential to the wearers through its mobile redesign studio offering a suitable space, tolos, and materials as well as skills and knowledge to those who can redesign their own garments. Design Stories similarly conducts workshops for instructing and developing the redesign processes together with others (partners and clients). Such workshops are conducted at many places, including colleges, companies, and organizations and cover several different topics, for example, design process, sustainable design, design and waste, energy, and design, among others. It further shares certain cases and project stories through its website. However, both the responsibility and the amount of information communicated by these brands are not as extensive as can be found in the case of the large slow fashion brands, heavily focused towards brand development. Instead these efforts are mainly small-scale and project-oriented. The other brands incorporated in the study merely communicate any further information on their post-retail initiatives through their websites apart from just portraying themselves as slow fashion.

Overall, it can be highlighted that the slow fashion brands exercise the highest degree of responsibility through their post-retail initiatives. This is attributed by more holistic physical management of their own used brand by undertaking multiple reverse logistics activities, along with extensive efforts taken in communicating sustainability. In fact these slow fashion brands epitomize their post-retail concepts and initiatives through social media, particularly their websites and stores, in such an explicit way that customers without the slightest hesitation can grasp their core business priority. Fast fashion brands and retailers and the small redesign brands follow next in terms of their total liability towards closing the loop. Although the fast fashion businesses do outsource most of the activities concerning physical management of the used items through partnership, they financially bear the cost of it. Furthermore, these fast fashion businesses engage in mega events and competitions to exemplify their efforts although a deeper look may raise issues related to transparency in information sharing along the revalue chain. The redesign brands undertake physical management of the items in a slightly different way. Even though these brands outsource most of the value appropriation activities such as washing, sewing, and the like, they take complete responsibility in upcycling leftovers and wastes of other actors to make something new and valuable. However, possibly due to financial limitations owing to the smaller size these brands have a lower level of communication efforts through various channels.

Table 2 summarizes the above discussion in terms of closing the material and liability loops and establishes a qualitative rating of the sustainable value-generation efforts for each business model category.

10 Brand Positioning: Are the Fast Fashion Businesses "Stuck in the Middle"?

With the increasing cults of throw-away fashion and hence growth of textile wastes, the post-retail segment offers business opportunities for fashion companies beyond just recyclers, charities, and second-hand retailers (Hvass 2014). The post-retail segment and businesses associated with it are nothing new and for several decades copious activities such as donation of used clothes to charities, reselling used clothes in flea markets, and so on have existed, even though the system of diverting wastes to landfill and/or energy recovery still maintains the majority. However, recently in light of the unsustainable practices of the textile and fashion industry resulting in depletion of the world's resources and environmental problems gaining prominence (Birtwistle and Moore 2007), fashion businesses are increasingly undertaking post-retail initiatives to overcome their unsustainable practices and image. Hvass (2014) highlights that in such an environment, fashion companies are increasingly rethinking their existing value propositions and collaborations with a multitude of stakeholders to devise sustainable solutions for closing the loop. The relation between sustainability and fashion business models in the case of forward value chains is pretty straightforward, concluding that the resource-hungry, cheap fast fashion businesses are unsustainable, whereas the slow fashion brands built upon timeless design, high durability, and green concepts are more sustainable in nature (Fletcher 2013). However, the relation is less explored when it comes to different post-retail initiatives undertaken by fast and slow fashion businesses.

Post-retail responsibilities for these fashion companies, in general, can be distinguished predominantly into two main categories, second-hand retailing and product take-back schemes, and are increasingly driven by the demands of generating a circular economy (Ellen McArthur Foundation 2013). The key environmental factors driving these operating post-retail business models aim towards generating a resource-efficient circular economy, and have been prioritized by all important actors in civil society, business, and government. These drivers fall back to the notion of closing the loop both by:

1. Recirculating the raw materials back into the consumption cycle in infinite loops
2. Taking extended liability or responsibility in handling the material reflow and related information

Although the first closing the loop process is conducted through reverse logistics activities following the five-R (i.e., reuse, reduce, recycle, redesign, and reimagine) approach to take care of the tangible flow, these additionally extend the intangible liabilities or responsibilities of the companies in managing the circularity.

In this context, traditional business model components, viz. key activities, customer value proposition, and economic profit formula (Magretta 2002), cannot alone define the success and value generated through these businesses. Sustainable value generation in this context involves closing both the material and liability loops to attain a resource-efficient circular economy and exercise extended organizational responsibility.

Although implementing a closed material loop through post-retail initiatives, it can be concluded that the fast fashion companies show minimum resource efficiency owing to their business strategy of offering discount vouchers to customers on donating used clothes. In such a case, the potential to displace the purchase of a new item is considerably low owing to the possibility that customers get induced into the spiral of throwing way old stuff from their wardrobe more frequently than ever to get discounts on new purchases. In fact such a spiral can increasingly lead to a scenario of perpetual discounts available year-round in the stores, a symbol of fast fashion culture (Bhardwaj and Fairhurst 2010; Birtwistle and Moore 2007). On the other hand, the small redesign brands offering no possibility to buy products made out of virgin raw materials can be considered to offer higher displacement potential.

Findings from the current study further show that in the liability loop, product stewardship executed by the fast fashion businesses is comparatively lower than the slow fashion brands'. Out of the two predominant business strategies highlighted by Hvass (2014), the fast fashion businesses only undertake product take-back schemes through their retail stores, whereas the slow fashion brands combine it with reselling as a part of their post-retail responsibility. Furthermore, some of the slow fashion brands follow a leasing business model thus taking complete ownership of the products throughout its extended lifetime thus increasing the extent of responsibility undertaken. In addition, lack of traceability and of communicating transparency regarding used clothes flow further decreases credibility for the collectors (fast fashion companies), hence their commitment in post-retail activities as perceived by the wearers (Palm et al. 2014). Even though there is an increasing need to communicate more information regarding the treatment of the collected clothes, not much has been done by the retailers to address this issue. With a higher degree of control on the post-retail initiatives the slow fashion brands possess and also communicate their efforts more transparently through social media and various other channels. Overall, the fast fashion businesses are positioned quite delicately in terms of generating their sustainable value in the post-retail initiative, as shown in Fig. 2.

The main challenge of the fast fashion business model in terms of the sustainability advantage in generating value in post-retail market, as depicted in Fig. 2, can be labeled the "stuck in the middle" challenge. On one hand the fast fashion businesses show a lower level of responsibility compared to that executed by the slow fashion brands, whereas in terms of displacement potential they can be rated the least.

Even though it emerged that the trade-off between sustainability and "speed of fashion" is not completely dichotomized in the case of post-retail initiatives,

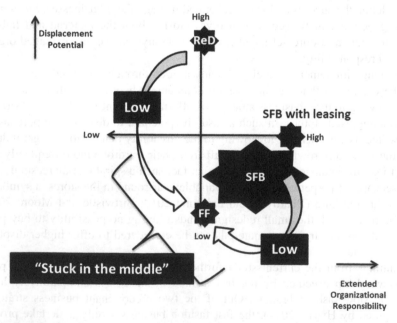

Fig. 2 Brand positioning in post-retail segment. *FF* Fast fashion businesses, *SFB* slow fashion brands, *ReD* slow redesign brands

and fast fashion businesses do execute considerable stewardship in "responsibility management," they are, however, under competitive pressure from both slow fashion and redesign brands in terms of closing the liability and material loops, respectively. In this context it could perhaps be concluded that the "stuck in the middle" positioning of the fast fashion companies may eventually reduce their sustainability advantage due to emerging pressures from both sides by slow fashion and redesign brands.

11 Conclusion

The purpose of this study was to explore whether the trade-off between sustainability and "speed of fashion" in fashion business models (classified into slow and fast fashions) as evident in forward value chains is equally observable in post-retail initiatives. In this context, this explorative study sheds light on various issues related to post-retail initiatives taken in the fashion industry in terms of sustainable value generation and strategic positioning of various operating fashion business models. First, a deeper understanding of diverse post-retail initiatives undertaken by various fashion businesses is presented, and is thematically analyzed into traditional business model components, viz. key activities, customer value proposition, and profit formula. The study categorized these fashion business models into four broad types: fast fashion, slow fashion, slow fashion with leasing, and redesign, based upon the differences underlying their business model components. A deeper conceptual underpinning to post-retail initiatives is provided in terms of closing the loop, viz. material efficiency and liability. Secondly, the study advances the understanding of the source of sustainable value generation in the post-retail market in terms of closing the material and liability loops, thus adding to environmental and social profit generation. This way it conceptualizes the success drivers of post-retail businesses. By analyzing these post-retail initiatives of various slow and fast fashion businesses, the study shows that the trade-off between sustainability and speed is not as rigid in the case of the reverse value chain as it is in the forward value chain, probably because there is always an aspect of greenness and social responsibility rendered by all businesses when operating in post-retail. However, the fast fashion businesses are somewhat in a "stuck in the middle" position in comparison to the slow and redesign brands along the material and liability loops.

Future research along this line can explore various possibilities. Quantitative and simulation studies can be conducted to explore the role and benefits of both displacement potential and product responsibility in closing the loops to attain higher resource efficiency and liability. These studies can delve in detail into many issues, including consumer purchase behavior with and without discounts, effect of redesigning on new purchases, effect of reuse on new purchases, and so on. Furthermore, a detailed quantitative formulation of the results of this study is highly desirable.

References

Armstrong CM, Niinimäki K, Kujala S, Karell E, Lang C (2015) Sustainable product-service systems for clothing: exploring consumer perceptions of consumption alternatives in finland. J Clean Prod 97:30–39

Bhardwaj V, Fairhurst A (2010) Fast fashion: response to changes in the fashion industry. Int Rev Retail Distrib Consumer Res 20(1):165–173

Birtwistle G, Moore CM (2007) Fashion clothing-where does it all end up? Int J Retail Distrib Manag 35:210–216

Bisgaard T, Henriksen K, Bjerre M (2012) Green business model innovation: conceptualisation, next practice and policy *Nordic Innovation Report*. Nordic Innovation Publication, Oslo

Bruce M, Daly L (2006) Buyer behaviour for fast fashion. J Fashion Market Manag 10(3):329–344

Carlsson J, Pal R, Mouwitz P, Lidström A (2014) ReDesign kläder: Förstudie [ReDesign Clothes: Prestudy]. The Swedish School of Textiles, Borås

Caro F, Martinez-de-Albeniz V (2014) Fast fashion: business model overview and research opportunities. In: Agrawal N, Smith SA (eds) Retail supply chain management: quantitative models and empirical studies, 2nd edn. Springer, New York

Choi TM (2013) Local sourcing and fashion quick response system: the impacts of carbon footprint tax. Transp Res Part E Logist Transp Rev 55:43–54

Christopher M, Lowson R, Peck H (2004) Creating agile supply chains in the fashion industry. Int J Retail Distrib Manag 32(8):367–376

Clark H (2008) SLOW + FASHION—an oxymoron—or a promise for the future …? Fashion Theory 12(4):427–446

Cline EL (2012) Overdressed: the shockingly high cost of cheap fashion. Portfolio/Penguin Group, New York

Cresswell JW (2007) Qualitative inquiry & research design: choosing among five approaches, 2nd edn. Sage, Thousand Oaks

Deloitte (2013) Fashioning sustainability 2013: redesigning the fashion business. In: Christiansen, AM, Hvidsteen K, Haghshenas B (eds) Deloitte

EC (2011) Communication from the Commission to the European Parliament, the Council, the European Economic and Social Committee and the Committee of the Regions: Roadmap to a Resource Efficient Europe. European Commission, Brussels

Eisenhardt KM, Graebner ME (2007) Theory building from cases: opportunities and challenges. Acad Manag J 50(1):25–32

Ekström K, Salomonsson N (2014) Reuse and recycling of clothing and textiles: a network approach. J Macromarket 1–17

Ekvall T, Watson D, Kiørboe N, Palm D, Texie H, Harris S, . . . Dahlbo H (2014) EPR systems and new business models: Reuse and recycling of textiles in the Nordic region. In: The Nordic Councils of Ministers (ed) TemaNord:Norden, Copenhagen

Ellen McArthur Foundation (2013) Towards the circular economy *Economic and Business Rationale for an Accelerated Transition* (vol 1). Ellen MacArthur Foundation

Farrant L, Olsen AI, Wangel A (2010) Environmental benefits from reused clothing. Int J Lifecyle Assess 15:726–736

Fleischmann M, Kuik R (2003) Production, manufacturing and logistics: on optimal inventory control with independent stochastic item returns. Eur J Oper Res 151(1):25–37

Fleischmann M, van Nunun JA, Grave B, Gapp R (2004) Reverse logistics: capturing value in the extended supply chain. ERIM Report: ERS-2004-091-LIS

Fletcher K (2010) Slow fashion: an invitation for systems change. Fashion Practice 2(2):259–266

Fletcher K (2013) Design for sustainability in fashion and textiles. In: Black S, de la Haye A, Entwistle J, Rocamora A, Root RA, Thomas H (eds) The handbook of fashion studies. Bloomsnury, London

Fletcher K, Grose L (2012) Fashion and sustainability: design for change. Laurence King Publishers, London

Gardetti MA, Torres AL (2013) Sustainability in fashion and textiles, values, design, production and consumption. Greenleaf Publishing, Sheffield

Guardian (2015) http://www.theguardian.com/sustainable-business/2015/aug/25/hms-1m-recycling-prize-clever-overproduction-fast-fashion. Accessed 01 Sep 2015

Ho HP, Choi TM (2012) A five-R analysis for sustainable fashion supply chain management in Hong Kong: a case analysis. J Fashion Market Manag 16(2):161–175

Hvass KK (2014) Post-retail responsibility of garments: a fashion industry perspective. Fashion Market Manag 18(4):413–430

Jayaraman V, Luo Y (2007) Creating competitive advantages through new value creation: a reverse logistics perspective. Acad Manag Perspect 21(2):56–73

Joy A, Sherry JF, Venkatesh A, Wang J, Chan R (2012) Fast fashion, sustainability, and the ethical appeal of luxury brands. Fashion Theory 16(3):273–296

Kelly M (2012) In the loop. Ecotextile News, Sweden, pp 44–46

Kibert NC (2003) Extended producer responsibility: a tool for achieving sustainable development. J Land Use 19(2):503–523

Kostecki M (1998) The durable use of consumer products: new options for business and consumption. Kluwer Academic Publishers, Dordrecht

Lindhqvist T (1992) Extended producer responsibility as a strategy to promote cleaner production. In: Paper presented at the proceedings of the invitational seminar, Trolleholm Castle, Sweden

Lindhqvist T (2000) Extended producer responsibility in cleaner production: policy principle to promote environmental improvements of product systems. Ph.D., Lund University, Lund

Lindhqvist T, Lidgren K (1990) Modeller för Förlängt producentansvar [Model for extended producer responsibility], pp 7–44. Ministry of the Environment, Från vaggan till graven-sex studier av varors miljöpåverkan

Lindhqvist T, Lifset R (1997) What's in a name: producer or product responsibility? J Ind Ecol 1(2):6–7

Magretta J (2002) Why business models matter. Harvard Bus Rev

Masson R, Iosif L, MacKerron G, Fernie J (2007) Managing complexity in agile global fashion industry supply chains. Int J Logistics Manag 18(2):238–254

Mattila H, King R, Ojala N (2002) Retail performance measures for seasonal fashion. J Fashion Market Manag 6(4):340–351

McKinsey (2011) Resource revolution: meeting the world's energy, materials, food, and water needs McKinsey and Company Sustainability, Resource Productivity Practice. McKinsey Global Institute. www.mckinsey.com/. Accessed 17 July 2015

Mont O, Dalhammar C, Jacobsson N (2006) A new business model for baby prams based on leasing and product remanufacturing. J Clean Prod 14(17):1509–1518

Myrorna (2015) http://myrorna.se/om-myrorna/sa-gar-det-till/. Accessed 14 Aug 2015

Niinimäki K, Hassi L (2011) Emerging design strategies in sustainable production and consumption of textiles and clothing. J Clean Prod 19:1876–1883

Niinimäki K, Pedersen E, Hvass KK, Svengren Holm L (2015) Fashion industry and new approaches for sustainability. In: Muthu SS (ed) Handbook of sustainable apparel production. CRC Press, Boca Raton

Nudie (2015a) http://www.nudiejeans.com/reuse/#/nudie-jeans-good-environmental-choice/. Accessed 14 Aug 2015

Nudie (2015b) http://www.nudiejeans.com/reuse/. Accessed 14 Aug 2015

Nudie (2015c) http://www.nudiejeans.com/recycle/. Accessed 14 Aug 2015

OECD (2001) Extended producer responsibility: a guidance manual for governments. OECD Publishing, Paris

Osterwalder A, Pigneur Y, Tucci C (2005) Clarifying business models: origins, present, and future of the concept. Commun Assoc Inf Syst 16(1). http://aisel.aisnet.org/cais/vol16/iss11/

Pacheco-Martins A, Pal R, Torstensson H (2014) Advanced computing techniques: new tools for fast fashion sales forecasting. In: Paper presented at the Ambience 2014, Tampere, Finland

Pal R (2015) EPR-systems and new business models for sustained value creation: a study of second-hand clothing networks in Sweden. In: Paper presented at the 15th AUTEX world textile conference, Bucharest, Romania

Palm D, Elander M, Watson D, Kiørboe N, Lyng K-A, Gíslason S (2014) Towards a new Nordic textile commitment: Collection, sorting, reuse and recycling. In: N. C. o. Ministers (ed) NORDEN Report, Copenhagen

Plank L, Rossi A, Staritz C (2012) Workers and social upgrading in 'fast fashion': the case of the apparel industry in Morocco and Romania. Austrian Foundation for Development Research (ÖFSE), Vienna

Pookulangara S, Shephard A (2014) The slow fashion process: rethinking strategy for fast fashion retailers. In: Choi TM (ed) Fast fashion systems: theories and applications. CRC Press, Leiden, pp 9–22

Re-dress (2015) http://www.re-dress.org.uk/index.html. Accessed 24 June 2015

Reuters (2012) http://www.reuters.com/article/2012/09/11/idUS56267+11-Sep-2012+HUG20120911. Accessed 25 Aug 2015

Serel DA (2014) Flexible procurement models for fast fashion retailers. In: Choi TM (ed) Fast fashion systems: theories and applications. CRC Press, Leiden, pp 59–75

Stahel W (1982) The product life factor. In: Orr GS (ed) An inquiry into the nature of sustainable societies. the role of the private sector, pp 72–105. http://infohouse.p102ric.org/ref/133/32217.pdf. Accessed 32220.32208.32015. Houston Area Research Centre, Houston

Stahel W (1994) The utilization-focused service economy: resource efficiency and product-life extension. In: Allenby B, Richard D (eds) The greening of industrial ecosystems. The National Academies

Stahel W (2007) Resource-miser business models. Int J Environ Technol Manag 7(5/6):483–495

Studio Re:design (2014) http://epi.vgregion.se/upload/Studio%20ReDesign/Rapporter/REDESIGN%20-%20FINAL.pdf. Accessed 20 Aug 2015

Sull D, Turconi S (2008) Fast fashion lessons. Bus Strat Rev 19(2):4–11

Tojo N (2004) Extended producer responsibility as a driver for design change: utopia or reality?. Lund University, Lund

Tukker A (2015) Product services for a resource-efficient and circular economy: a review. J Clean Prod 97:76–91

Wang K, Gou Q, Yang L, Siqing S (2014) Coordination of a fast fashion supply chain with profit-loss sharing contract. In: Choi TM (ed) Fast fashion systems: theories and applications. CRC Press, Leiden, pp 77–92

Wikipedia (2015) Slow Fashion, https://en.wikipedia.org/wiki/Slow_Movement#Slow_Fashion. Accessed 10 Aug 2015

WRAP (2013a) Study into consumer second-hand shopping behaviour to identify the re-use displacement effect. Project Code MDP007-001

Yunus M, Moingeon B, Lehmann-Ortega L (2010) Building social business models: lessons from the Grameen experience. Long Range Plan 43:308–325

Hanji, the Mulberry Paper Yarn, Rejuvenates Nature and the Sustainable Fashion Industry of Korea

Kyung Eun Lee and Eulanda A. Sanders

Abstract Sustainable production and consumption in the fashion industry has become an imminent and crucial phenomenon globally. Adoption of natural fibers is an eco-design approach to accomplish sustainable product development capitalizing on environmentally conscious characteristics such as renewability and biodegradability. Therefore, several Korean textile companies (e.g., Ssang-Young, Oh-Sung) and textile research institutions (e.g., Korea Institute for Knit Industry) have led the development of Hanji yarn, a yarn made from cellulosic fibers of the inner bark of the mulberry paper trees (IBMP) using manufacturing processes adapted from traditional Korean paper production methods. The unique raw materials and novel production processes used to produce Hanji yarn has resulted in a textile product that can increase the aesthetics and functionality of apparel and products of other product categories. Fabrics created from Hanji yarn have been adopted by several global apparel companies, specifically within the activewear market. The success of Hanji yarn within the apparel and textile industry is a prime example of integrating sustainable materials and technology to promote green fashion in the global market.

Keywords Sustainability · Hanji yarn · Natural materials · Corporate social responsibility

K.E. Lee (✉)
Department of Apparel, Events and Hospitality Management,
Iowa State University, 28 Mackay Hall, Ames, IA, 50011, USA
e-mail: Kyungeun@iastate.edu

E.A. Sanders
Department of Apparel, Events and Hospitality Management,
Iowa State University, 1052 LeBaron, Ames, IA, 50011, USA
e-mail: Sanderse@iastate.edu

© Springer Science+Business Media Singapore 2016
S.S. Muthu and M.A. Gardetti (eds.), *Green Fashion*,
Environmental Footprints and Eco-design of Products and Processes,
DOI 10.1007/978-981-10-0111-6_7

159

1 Introduction

1.1 Definition of Sustainability

Sustainability is a fundamental process of businesses to protect the environment and prevent harm to natural resources (Khan and Islam 2015; Choudhury 2015). Sustainability in business concerns an entire life cycle of natural resources, how these resources are collected, processed, applied, and refilled, and the influence of consumption and discarding of manufactured products (Khan and Islam 2015). Gardetti and Muthu (2015) explain sustainability from an entrepreneurial standpoint, which may embrace pollution avoidance, efficient resource utilization, adoption of sustainable technology, corporate social responsibility (CSR), human rights, transparent management, and stakeholder relationship management (SRM). Defining sustainability in relation to apparel manufacturing, Roy Choudhury (2015) addresses the importance of balancing three areas within the supply chain: (a) society (e.g., fair trade, human rights), (b) economy (e.g., economic returns), and (c) environment (e.g., efficient energy use, recycling raw materials and energy). Sustainability of apparel products refers to creating an environmentally and socially responsible design philosophy and trend in products (Adams and Frost 2008). In accordance with the definitions in previous studies related to sustainability, this case study is focused on how apparel products are made from natural materials.

1.2 Impact of Apparel Industry on Environment

The apparel industry comprises a large sector of the world's economy and contributes approximately US $3 trillion (Gardetti and Muthu 2015). The harmful impact of the apparel industry to our environment, and ultimately our society, occurs in the various stages of the apparel product life cycle: through the production of raw materials (fibers, yarns, and textiles), production of garments (construction, packaging, and logistics), and post-production activities related to the end-use (recycling or discarding) of products (Khan and Islam 2015). Problematic sustainability issues in the apparel industry result from the development of textiles involving the use of toxic chemicals, synthetic materials, and energy in wet-processes such as dyeing, printing, and finishing (Sivaramakrishnan 2009). The accumulation of harmful synthetic chemical residuals used in textile production is discharged directly into water sources, negatively affecting the soil, water, and environment (Oceotextiles N.D.). There are over 70,000 synthetic chemicals used commercially, despite a lack of scientific testing on their potential impact on the environment (Sivaramakrishnan 2009). In addition to chemical discharges, an excessive amount of water and electrical energy is consumed during textile manufacturing processes (Oceotextiles 2013). The stagnation of technological developments and lack of further concrete research in the textile industry affects our environment because textile manufacturers usually depend on their previous knowledge and experiences that are kept confidential rather than scientific approaches (Roy Choudhury 2015).

1.3 What Is a Sustainable Firm?

The Sustainable Apparel Coalition (SAC, N.D.) defines the sustainable apparel firms as the ones that make an effort to aggressively reduce environmental impact of their products such as Nike, Patagonia, Stella McCartney, Eileen Fisher, and NAU. For instance, Patagonia is a member of the Blue Sign® system, which is one of the most respectable environmental sustainability auditing systems in textile production from fiber to processes, and the company produced its 100 % base layer products solely with Blue Sign® system approved fabrics (Fetcher 2012). Organizations, developed countries, environmental protection agencies, and the apparel companies have developed production standards and regulations to control the disposal of toxic materials such as the Oeko-Tex standard, Blue Sign, and the Global Organic Textile Standard (Abreu 2015). Previous studies have identified three conceptual approaches characterizing a firm's endeavors to protect the environment: (a) corporate social responsibility (CSR), (b) green supply chain management (GSCM), and (c) eco-design. Khan and Islam (2015) assert that the importance of an expanded role of the designer, as well as other stakeholders such as the manufacturer, merchandiser, and consumer, is needed in the production, usage, and disposal of the apparel products in consideration of current environmental issues (see Tables 1 and 2).

Table 1 Hanji yarn's global market growth (Korea Institute for Knit Industry 2015) (Unit: US $1 million)

Year	2010	2011	2012	2013	2018 (Expected)
Hanji paper	0.15	0.49	0.86	1.27	11.82
Hanji yarn	0.15	1.82	3.18	4.55	54.55
Hanji fabric (Knit)	1.73	6.18	10.91	15.45	109.09
Hanji textile products	25	25	43.64	61.82	436.36

Table 2 Hanji yarn production capacity growth in Korea (Korea Institute for Knit Industry 2015) (Unit: ton)

Year	2011	2012	2013	2018 (Expected)
Hanji paper	100	200	300	750
Hanji yarn	220	380	550	1,400
Hanji fabric (Knit)	600	1,000	1,500	3,700

1.3.1 Corporate Social Responsibility

The commitment of all stakeholders in the apparel companies to sustainable practices is demonstrated in complying with the corporate social responsibility programs. CSR is a firm's obligation to execute their business activities in a manner that is not destructive to society or to the environment (Steiner and Steiner 2009). Apparel companies need to develop initiatives to motivate stakeholders including the owners, members along the supply chains, and retailers, to be dedicated to green-fashion businesses (Abreu 2015).

1.3.2 Green Supply Chain Management

Green supply chain management (GSCM) is a conceptual approach used to enhance the economic positioning of a firm's global environmental initiatives (Zhu et al. 2005). GSCM creates awareness in apparel companies in order to direct their businesses towards eco-friendly approaches, which are advantageous both to protect the environment and to use sustainability to enhance the brand image of a company (Zhu et al. 2005). ISO 14001 is one of the global institutional certifications used to acknowledge a company's performance in satisfying GSCM environmental requirements. ISO 14001 regulations measure a firm's environmental management system in legal terms, as well as other requirements (International Organization for Standardization 2004). Global environmental certifications provide a systemized method to evaluate an apparel company's level of commitment to eco-friendly standards.

1.3.3 Eco-design

Apparel companies should initiate a phase within product design with sustainable practices in mind (Choudhury 2015). Eco-design is creating products with an emphasis on the environmental effects and responsibilities. Therefore, all decisions and actions during design and product development processes adhere to environmental approaches (Lewis et al. 2001). Eco-design can enable a firm's long-term growth by increasing innovative product designs, in which the product's core element is sustainability (Pigosso et al. 2013). In addition, eco-design generates competitive new business opportunities and paradigms, allowing additional perceived value to the companies (Hansmann and Claudia 2001).

1.4 Natural Materials and Hanji Yarn

Sustainable product design may be accomplished by various methods, such as selecting natural fibers and processing methods, building sustainable consumption behaviors, and recycling materials (Khan and Islam 2015). An example of a sustainable design approach is when apparel designers adopt natural fibers and dyeing methods, absent insecticides or fertilizers. Utilization of organic, recycled, and naturally dyed fibers are more favorable to the environment as these fibers are renewable and biodegradable (Choudhury 2015). Natural cellulosic fibers such as cotton, flax, hemp, jute, mulberry, and ramie have been adopted in the apparel industry for their multifunctionalities, such as ventilation, antimicrobial, moisture absorption, and wicking properties (Choi et al. 2012; Jang et al. 2015; Jeong et al. 2014; Kim 2006). Fibers created from mulberry have characteristics of cellulosic fiber, which have superior breathability, ventilation, moisture absorption, wet strength, and antimicrobial properties (Choi et al. 2012; Jeong et al. 2014; Kim 2006). Mulberry is one of the most environmentally sustainable materials because of its rapid growth and high germinative power, especially in poor soil and climate challenges, similar to the cases of bamboo (Kew Royal Botanical Gardens N.D.; Saito et al. 2009; Xu et al. 2011). By-products from the mulberry tree are widely used as an ingredient for various products across numerous industries, such as papermaking, food, optical, pharmaceutical, and textile industries, due to its strong sustainability and multifunctionalities (Jang et al. 2015). Hanji yarn is made with mulberry paper, which is produced in the adapted Korean traditional paper-making techniques: molding, layering, and burnishing to allow strength and flexibility (Choi et al. 2012; Kim 2006). The mulberry paper is transformed into yarn through the processes of slitting, twisting, and weaving/knitting that generate minimal negative environmental impacts (Ssang-Young 2015).

1.5 The Purpose of the Case Study

Despite Hanji yarn's sustainable attributes discussed earlier and its rapid increase of revenue growth in the global market, there is a lack of literature describing uses of Hanji yarn in the fashion industry. The purpose of this case study is to: (a) review the manufacturing processes and sustainable characteristics of Hanji yarn, (b) identify the current business state of Hanji yarn in the fashion industry, and (c) suggest directions for future research for the natural fibers to contribute to green fashion business practices. Additionally, this chapter provides information on Hanji yarn to stakeholders in the apparel industry: including designers, manufacturers, merchandisers, and consumers.

This study was conducted based on a comprehensive review of previous studies about Hanji yarn and seven documents provided by six organizations that use Hanji yarn in their businesses in Korea and the United States: Hanji yarn manufacturers (Ssang-Young), branded labels (Hyundai Dymos; Isae; The North Face Korea; Troa), and a textile research and development laboratory (Korea Institute for Knit Industry). These documents are in a variety of formats determined by each company. For example, Ssang-Young presented the company's comprehensive analysis report of their Hanji yarn business from the years 2007 to 2015. Isae and Troa provided the company's press kits, containing images of the Hanji yarn manufacturing processes, images of the products, and visual presentation of retail locations selling these products made with the Hanji yarn. The North Face Korea shared the manufacturing processes of Hanji yarn T-shirts and product images. Hyundai Dymos reported a conceptual car-seat cover developed from Hanji paper and yarn in 2015. The Korea Institute for Knit Industry provided a Hanji textile product market analysis report written in 2015 and the project report of the Warm-Biz apparel line development using Hanji yarn executed in 2014.

2 Use of Mulberry Products

The pulp fiber from mulberry trees has traditionally been used as a raw material to make papers mainly in Asian countries such as Korea, Japan, China, Thailand, and the Philippines (Jang et al. 2015). Hanji is a Korean paper made with mulberry pulp using traditional Korean paper-making techniques such as molding, layering, and burnishing, which originated over 1,600 years ago (Choi et al. 2012; Kim 2006). Hanji yarn was developed by Korean manufacturers and research institutions starting in the late twenty-first century (E-daily News 2013; Hankook Daily 2009). The original Hanji in a nonwoven paper structure is transformed into yarn through modern production techniques such as paper slitting/twisting and water-supplying (Ssang-Young 2015). The combination of the unique fiber structure and novel production processes allows Hanji yarn to have excellent breathability and strength, which enhances wearing comfort (Choi et al. 2012; Kim 2006).

Hanji yarn can be woven and knitted from low to high yarn counts and can be blended with other cellulosic fibers such as cotton and rayon to enhance tensile strength and morphostasis of the fabric (Korea Institute for Knit Industry 2015; Park and Lee 2013). Hanji yarn's stiffness and rough surface prohibit a stable and uniform twist of the yarn, which can result in yarn breakage and needle disruption during the knitting processes. Because of this, the mulberry paper strips used to make the yarn have been redeveloped as a ply-twisted or filament yarn (Park and Joo 2012).

3 How Hanji Yarn Is Manufactured

3.1 Historical Footnotes of Hanji Paper

The Goryeo Dynasty of Korea (918–1392) was the most flourishing period of printed publications using Hanji paper (Yum et al. 2009). Hanji paper-making techniques were developed in the eighth century (Yum et al. 2009). Mulberry pulp was the most commonly used raw material for paper-making (Kim 2006). Hanji was originally used as a material to cover walls, ceilings, windows, and doors in traditional Korean housing due to its natural breathable properties (Kim 2006).

The durability of Hanji paper has been proven by the condition of nine publications printed on Hanji from the fourteenth century to the early twentieth century. These documents are listed on the Memory of the World Register by the United Nations Educational, Scientific, and Cultural Organization (UNESCO) and are all in good condition (UNESCO N.D.). Previous studies show the incorporation of technologies to the Hanji paper properties, such as titania nanorods and gamma irradiation. Applying titania nanorods to the surface of mulberry fibers enhances antimicrobial, antiyellowing, and self-cleaning capabilities by spreading UV radiation through titania nanorods (Jang et al. 2015). Gamma irradiation of the Hanji paper improves the sterilization of organic products such as insects and fungi (Choi et al. 2012). The researchers' efforts in development of Hanji paper preservation methods using advanced technologies allow new opportunities for textile manufacturers to produce durable Hanji yarn (Korea Institute for Knit Industry 2015).

3.2 Raw Materials of Hanji Yarn

The mulberry paper tree has many properties advantageous to the paper-making industry, such as high productivity, ease of cultivation, and usefulness of biomass resources (Takasaki et al. 2011). The raw material of Hanji paper is a pulp fiber from the inner bark of the mulberry paper tree (IBMP) that is made of a fibroid material with numerous minute gaps that allow air circulation. Thus, it possesses great ventilation and humidity control (Choi et al. 2012; Kim 2006). The Korea Institute for Knit Industry (2015) explains that the pulp fiber of IBMP has a long length and narrow width, which results in higher tensile strength, compared to short fiber lengths in other yarns. The average fiber length of the IBPM is the second longest at 9.37 mm, behind hemp at 14.46 mm and IBPM's average width is 0.027 mm (Korea Institute for Knit Industry 2015). The chemical composition of IBMP consists of 16.3 % of lignin, 69.1 % of holocellulose, and 13.7 % of pentosan; all substrates combined produce fibers with high moisture absorption, wet-strength, and antimicrobial properties (Jeong et al. 2014). Normally, the one-year-old mulberry paper tree is ideal to produce good quality paper, because it contains a high polymerization (DP) degree of 7,000–9,000 that creates uniform distributions of the fiber (Jeong et al. 2014).

3.3 Hanji Yarn Manufacturing Processes

The conventional method for manufacturing Hanji yarn consists of four steps: (a) Hanji paper is made by traditional Korean paper-making techniques of molding, layering, and burnishing, (b) the paper is cut into very fine strips in a designated width to form a paper tape, (c) the paper tape is twisted using a spinning wheel, and (d) the twisted paper tape is woven or knit into a fabric using a power loom

(Park and Lee 2013; Troa N.D.). There have been limitations to mass production of Hanji yarn by using conventional processes and old knitting instruments, such as unevenness in paper slitting and yarn breakage during the twisting processes. Thus, new manufacturing methods have been developed by researchers and manufacturers in four common processes to produce Hanji yarn: (a) the use of reinforcing agents in the Hanji paper-making process, (b) the cutting of Hanji tape with a rotary slitter, (c) the use of a continuous water-supplying instrument while twisting the Hanji tape into yarn, and (d) weaving to the target density and uniform structure (Park and Lee 2013).

3.3.1 Paper-Making Process

The distinctive processes in this stage are an adaptation of various raw materials and paper-reinforcing mediums from the early stage of paper making to overcome limitations of traditional methods (Park and Lee 2013). Paper yarns possess some disadvantages in flexibility in comparison to nonpaper-based yarns (Park and Joo 2012). Hence, a longer materials arranged in a parallel format should be selected to create a fine and strong Hanji yarn (Park and Joo 2012). Hanji paper is made at a neutral pH level (7.5 ~ 9.0) without using additional acids or stabilizing agents, therefore, it has a long lifespan and high preservability (Korea Institute for Knit Industry 2015).

3.3.2 Preparation of Hanji Tape Yarn

The prepared pulp fiber is processed to make a batter mixed with the reinforcing agents. Then, a finished batter is beaten repeatedly to generate the shape of paper. After that, the processed paper is cut at a predetermined width using a rotary slitter to form a paper tape (Oh et al. 2010; Park and Joo 2012). The Korea Institute for Knit Industry (2015) provided the scanning electron microscope (SEM) images in Fig. 1, comparing the fiber structure of Hanji 20s and cotton 30s yarn. Despite the fact that both yarns' fiber radii are 240 μm, Hanji yarn's fibers were agglomerated, whereas cotton yarn has a dispersed fiber structure. Hanji yarn's fiber agglomeration creates a high elastic resilience.

Hanji tape's weight and width determine the fineness of Hanji yarn. Tape yarns in lower weights and narrower widths result in finer yarns. General Hanji tape's weight is roughly 12–14 g/m^2, and its width is between 1.2–0.8 mm (Oh et al. 2010; Park and Joo 2012). The North Face Korea cut their Hanji tape in 1.5 mm to make a thicker, stronger yarn for outdoor sportswear (The North Face Korea 2015). Park and Lee (2013) suggest decreasing the weight of Hanji paper to around 8 to 10 g/m^2 to create a finer Hanji yarn. Park and Joo (2012) address the necessity of using a tape within the minimum weight of 13 g/m^2 to generate uniform distribution of the yarn's twisted structure. If a paper tape is less than 10 g/m^2 in weight,

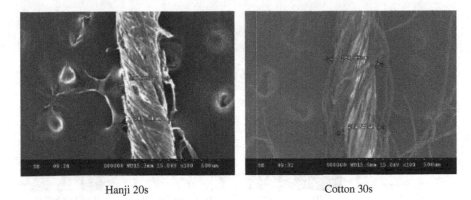

Hanji 20s   Cotton 30s

Fig. 1 Scanning electron microscope (SEM) photographs of Hanji and cotton yarn in 240-μm fiber radius (Korea Institute for Knit Industry 2015)

potential risks such as crumbling, irregular structures, and breakoff can occur during the slitting procedure (Park and Joo 2012).

3.3.3 Preparation of Paper Yarn

A finished Hanji tape is placed in the machine direction and inserted into a twisting cone-shaped device connected to a water supply, making the yarn (Oh et al. 2010; Park and Joo 2012; Park and Lee 2013). The water supply is set to feed water prior to twisting at a regular speed of 60–90 cc/min to supply water targeting the middle part of the paper yarn for major benefits: (a) improved breakoff, (b) increased flexibility, and (c) creating a uniform twist with less fluffiness than other natural yarns (Park and Joo 2012; Park and Lee 2013). Controlling cut-off times of the water supply influences the fineness of the yarn (Park and Lee 2013). The yarn count of woven textiles made from Hanji is often lower than textiles woven from cotton and hemp fibers, therefore Hanji textiles are more flexible in comparison to cotton and hemp textiles (Park and Joo 2012). See Fig. 2.

3.3.4 Weaving/Knitting Process of Paper Yarn

Hanji yarn allows both weaving and knitting to make textiles. The twisted Hanji yarn is woven to make a fabric in a 2 × 2 basket structured instrument (Oh et al. 2010). It is important to have a lightweight and uniformed distribution of mulberry papers to produce a very fine Hanji yarn (Oh et al. 2010; Park and Joo 2012). However, an overly uniform distribution may cause poor flexibility in Hanji yarn (Park and Lee 2013). Park and Lee (2013) found that the tensile

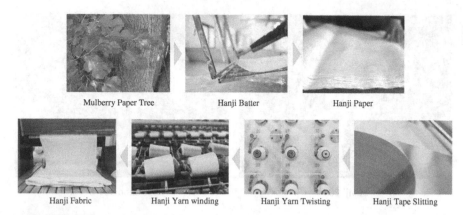

Mulberry Paper Tree   Hanji Batter   Hanji Paper

Hanji Fabric   Hanji Yarn winding   Hanji Yarn Twisting   Hanji Tape Slitting

Fig. 2 Hanji fabric production process (Troaco.com N.D.)

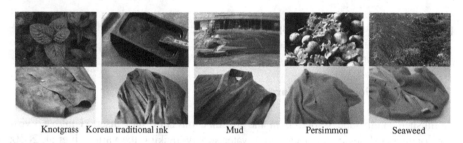

Knotgrass   Korean traditional ink   Mud   Persimmon   Seaweed

Fig. 3 Hanji fabric dyed in natural materials (Isae.co.kr N.D.)

strength of the Hanji fabric is higher in the warp direction than in the weft direction, based on the fabric density of $52 \times 40/in^2$.

3.3.5 Natural Dyeing

The Hanji fabric's excellent dyeing properties and color resolution capabilities are aptly suited to the natural dyeing techniques applied in Korean traditional methods (Isae 2015). Hanji can be dyed in a wide variety of natural materials. For example, Isae (2015) presented various natural dyeing methods using knotgrass (indigo), Korean traditional ink, mud, persimmon, and seaweed applied in different natural fabrics including Hanji. Troa (2015) sold their denim collection natural-dyed with knotgrass (indigo), Korean traditional ink, safflower, coal, and chestnut. See Fig. 3.

4 Advantages for Hanji Yarn

4.1 For the Environment

Hanji yarn is completely biodegradable, renewable, and recyclable as a 100 % natural material (Park and Joo 2012). Mulberry paper can be easily blended with other natural materials without compromising its desired characteristics. This promotes the use of natural materials to develop functional textiles (Korea Institute for Knit Industry 2015). Hanji yarn products dyed with natural materials do not generate toxic chemicals in water waste for manufacturing (Troa 2015). Hanji yarn provides a high level of ventilation and moisture absorption, which can be used to develop cooling clothing. Such clothing could contribute to reducing energy usage during summer months (Korea Institute for Knit Industry 2015).

4.2 For Consumers

There are four key advantages of Hanji yarn identified for consumers: (a) antimicrobial activity, (b) wearing comfort, (c) promotion of health, and (d) textile aesthetics. Such advantages of Hanjji yarn extend the range of its use to apparel, bedding, filtering, and hygienic products (Gil et al. 2010).

4.2.1 Antimicrobial Activity

The chemical composition of the inner bark of mulberry paper (IBMP) trees allows antimicrobial activity in Hanji fabric which helps to prevent atopic dermatitis (Ju et al. 2013) and provides deodorization (Jang et al. 2015; Park and Joo 2012; Takasaki et al. 2011). Ssang-Young (2015) compared antimicrobial activity between 100 % Hanji and 100 % cotton yarn by leaving both yarns at 77 °F with 80 % humidity for one month in April. The results of testing demonstrated that Hanji yarn was minimally damaged, whereas cotton was excessively decomposed (Ssang-Young 2015; see Fig. 4).

4.2.2 Wearing Comfort

The natural characteristics of IBMP in moisture absorption and wicking enhance the wearing comfort of Hanji fabric (Jang et al. 2015). The weight of Hanji fabric is only 0.5 g/cm^2, thus, Hanji yarn is favorable to blend with other materials without substantially increasing additional weight (Oh et al. 2010). Mulberry paper's high human skin affinity minimizes allergic or toxic effects to the body (Gil et al. 2010). Hanji yarn is categorized as a "filament yarn without mow," which has

Hanji yarn   Cotton yarn

Fig. 4 Ssang-Young's antimicrobial activity testing results comparing Hanji and cotton yarn left at 77 °F with 80 % humidity for one month in April (Ssang-Young 2015)

unique textures allowing fabric softness and a cooling sensation (Gil et al. 2010; Korea Institute for Knit Industry 2015; Park and Joo 2012).

4.2.3 Promotion of Health

Hanji yarn's high cellulose composition contains strong UV radiation-absorbing properties; hence, it effectively protects the skin from high-degree UV exposure (Gil et al. 2010). Mulberry paper has capabilities in far-infrared radiation in which the action boosts blood circulation as much as red clay (Gil et al. 2010; Ju et al. 2013). The potential health benefits of textiles made from Hanji yarn is an area of needed study by health and textile scientists.

4.2.4 Textile Aesthetics

The mulberry tree grown in Korea has a long fiber length and high durability because of a broad daily temperature range and four seasons (Gil et al. 2010). Park and Lee (2013) examined Hanji yarn's high tensile strength, which allows durability and resiliency of Hanji textile products. They also found that its fabric's color fastness in washing and dry-cleaning is 4 grade, and the stain-resistance is 4–5 grade in which both grades are equivalent to cotton fabric. Hanji yarn has dyeing properties that bring good color resolution (Gil et al. 2010; Korea Institute for Knit Industry 2015).

4.3 For Apparel Companies

Hanji yarn has been used to manufacture various sustainable products, such as apparel and home textiles in the fashion industry of Korea. Thanks to the continuous efforts of Hanji yarn manufacturers and governmental support, the global market for Hanji yarn has grown constantly over the last five years and Hanji yarn has become one of the premium fashion products in the global market (Korea Institute for Knit Industry 2015; MBC 2014).

The local economies of the textile industry in Korea have been improved due to the increasing volume of Hanji yarn production (MBC 2014). For instance, the Ssang-Young company made US $1.8 million by launching the Hanji yarn business in 2009 (Hankook Daily 2009) and grew its Hanji yarn's annual sales to US $2.7 million in 2013 (E-daily News 2013). Oh-Sung made a Hanji yarn procurement contract for US $2.7 million with Kolon Sportswear company in 2010 (Yonhap News 2010).

5 Current Business State of Hanji Yarn in the Fashion Industry

Mulberry-based fibers have been highly preferred in the textile industry especially for manufacturing of organic fabrics (Park and Joo 2012). Currently, the global market size of Hanji yarn products in the fashion industry is about US $50 million and estimated to be increased by US $500 million over the next five years (MBC 2014). Similar to the Hanji yarn business of Korea, Japanese paper yarn sales have grown over 10 times compared to 5 years ago in the global market (Ssang-Young 2015). Therefore, the global paper yarn market in general might be forecasted to grow rapidly in the future. Hanji yarn production contributes to the revitalization of local economies by generating new business and employment opportunities. For instance, Yonhap News (2010) reports that Northern Jeolla province in Korea was the hub of the textile industry with three of the largest textile manufacturers of the country in the 1980s and 1990s. However, from the late 1990s to mid 2000s, the textile business of Jeolla has dramatically decreased because of changes in market conditions and consumers' tastes. Successful Hanji yarn textile manufacturing business rejuvenated Jeolla's economy after its approximately 10 years' stagnation, helping over 8,000 employees in about 870 textile companies (Northen Jeolla Province Daily 2014). There are many global apparel companies (e.g., manufacturers, branded companies, research institutions), that have made distinctive contributions in green design and management with the adoption of Hanji yarn.

Fig. 5 Hanji yarn products: fabrics, yarns, socks, undergarments, and golf shirts (Lohashanji.com N.D.; Ssang-Young 2015)

5.1 Manufacturers

The producers of Hanji yarn include Ssang-Young, Oh-Sung, and Sung-Sil located in Jeolla province of Korea, which is the center of Hanji production. Ssang-Young's production capability of Hanji yarn is 10 tons per month (Ssang-Young 2015). Ssang-Young manufactures products in sportswear, undergarments, infantwear, ties, scarves, socks, and bedding textiles for their private labels and OEM (Original Equipment Manufacturer) contracts with apparel brands including sportswear companies such as The North Face, Millet, and Fila (Ssang-Young 2015). Ssang-Young holds two global certifications that prove their product's quality and sustainable management: ISO 9001:2008 (Quality Management) and ISO 14001:2004 (Environmental Management Systems; Ssang-Young 2015). Sung-Sil specializes in functional sock production (Yonhap News 2010). Oh-Sung produces apparel lines for sportswear and undergarment companies such as Kolon Outdoor sportswear and Wacoal Lingerie (Yonhap News 2010). See Fig. 5.

5.2 Research Institution (Korea Institute for Knit Industry)

Korea Institute for Knit Industry is a textile research institution specializing in R&D of apparel and industrial textile products, established in 2001. Korea Institute for Knit Industry has been actively involved in the development of eco-friendly fabrics using natural materials including Hanji yarn (Korea Fashion and Textile News 2015a, b). Warm-Biz is a government-funded project that Korea Institute for Knit Industry performed in 2015 in collaboration with the textile manufacturer Oh-Sung to develop a workwear line using different cellulosic fibers including Hanji yarn with lightweight, enhanced insulation, and moisture absorption functionalities (Korea Fashion and Textile News 2015a, b). Warm-Biz is a campaign corresponding to climate change to reduce energy consumption and CO_2 emissions that followed the success of Cool-Biz, which was initiated by the Japanese government in 2005 (Aliagha and Cin 2013). The Japanese government aggressively promoted the Cool-Biz and Warm-Biz campaigns to keep thermostats at 68 °F, while keeping less or extra layering to maintain the thermal comfort of clothing in 2011 (Aliagha and Cin 2013; Japan Today 2014). According to Korea Institute for Knit Industry's Warm-Biz project report (2015), the Warm-Biz Hanji fabrics were invented in both knitting and weaving techniques. The report (Korea Institute for Knit Industry 2015) illustrates the detailed information about the textiles and garments developed. The Hanji fiber of the knit fabric used in the innerwear line was blended with Hanji, cotton 24s, and spandex of differentiated ply twist and fineness (denier) to improve Hanji's elongation. The woven fabrics were developed in Hanji 30s and 13s yarn blended with cotton, in which cotton was applied in the warp direction and Hanji was implemented in the weft direction. After that, silver or gold dust was coated, and microperforated aluminum was laminated on the yarn to prevent releasing of body heat and reduce ammonia activity by 80 %. The Warm-Biz Hanji fabrics in the dobby texture were used to make a shirt and pants, and the fabrics in twill weave were applied to a jacket and pants. The Korea Institute for Knit Industry (2015) demonstrated the lab testing results comparing a body temperature conservation between a regular Hanji and silver dust coated Warm-Biz fabric in the images taken by a thermographic camera (see Fig. 6). The results presented that the silver-coated Warm-Biz Hanji fabric retained 84.2 °F, approximately 4 °F higher than the 80.6 °F of the regular Hanji fabric.

The Warm-Biz workwear line (see Fig. 7) developed by Korea Institute for Knit Industry will be distributed to government employees and expanded to company

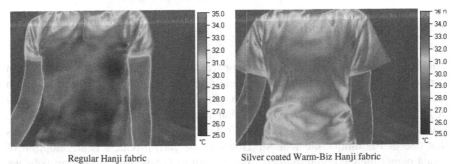

Regular Hanji fabric   Silver coated Warm-Biz Hanji fabric

Fig. 6 The thermographic pictures comparing body temperature conservation between the regular Hanji and silver dust coated Warm-Biz Hanji fabric (Korea Institute for Knit Industry 2015)

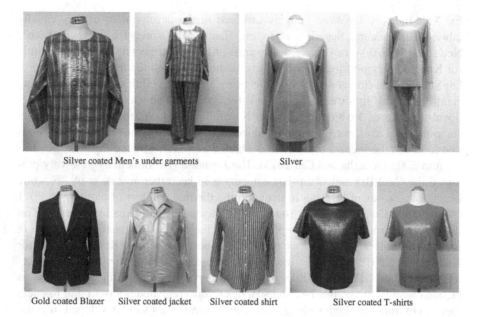

Silver coated Men's under garments   Silver

Gold coated Blazer   Silver coated jacket   Silver coated shirt   Silver coated T-shirts

Fig. 7 The Warm-Biz workwear line developed by Korea Institute for Knit Industry (Korea Institute for Knit Industry 2015)

workers starting in the winter of 2015 (Korea Fashion and Textile News 2015a, b). Korea Institute for Knit Industry developed another Hanji composite yarn funded by the Korean government in collaboration with the Dytec textile laboratory and the textile manufacturer Cotton Queen in 2015 (Northern Jeolla Province Daily 2014). Hanji composite yarn has enhanced functionalities in 8 % elongation and 80 % elastic resilience, which can be used to produce casual and outdoor sportswear apparels, compared to previously developed Hanji yarns (Korea Fashion and Textile News 2015a, b).

5.3 Branded Companies

Troa is a New York based eco-friendly apparel company that sold its branded Hanji yarn denim lines to leading fashion retailers in fashion capitals of the world such as New York, Tokyo, and Paris (Troa 2015). According to Troa's press kit (Troa 2015), buyers of Troa include top global retailers such as Barneys New York, Harvey Nichols, Ten Corso Como, and Collette. Troa's materials also demonstrates that one of the unique elements of Troa's Hanji yarn products is an adoption of abaca plants blended with mulberry paper yarn to create fabrics that are soft, yet more durable than cotton (Troaco.com N.D.). Of Troa's denim lines, 30–100 % are dyed naturally to create unique colors depending on the seasons, through the use

Troa Hanji denim lines and design director, Han Song

Fig. 8 Troa's Hanji products (Troa 2015)

of different dyeing techniques such as yarn dye and piece dye (Troa 2015). Troa sold "black tuxedo jeans" made with Hanji yarn dyed in Korean ink in collaboration with actress Gwyneth Paltrow and the online boutique Goop.com (Troa 2015). Troa continuously invests efforts in developing novel Hanji fabrics with high quality to produce their denim lines (Troaco.com N.D.). See Fig. 8.

Isae, founded in 2000, is a premium designer clothing and home textile label specializing in handmade natural fabrics dyed with natural materials (Isae.co.kr N.D.). Isae presents outerwear jackets, dresses, and vests in Hanji fabric dyed in

Fig. 9 Isae's Hanji apparel collection and flagship store (Isae.co.kr N.D; Isae 2015)

knotgrass (indigo), Korean traditional ink, mud, and persimmon. Their products
are sold in over 70 retail locations in Korea (Isae 2015). Isae produces most of
their products by hand in collaboration with Hanji yarn artisans (Isae.co.kr N.D.).
Various natural materials are used such as organic cotton, linen, ramie, hemp, and
kenaf for Isae's apparel and home textile products (Isae.co.kr N.D.). There are four
high-end conceptual flagship stores run by Isae, which contain clothing and acces-
sory lines, home textiles, and pottery, representing the brand image and lifestyle of
the company (Isae 2015; see Fig. 9). The flagship stores target international consumers who visit Korea to test Isae's products for preparing additional store open-
ings in global locations (Isae 2015).

The North Face Korea initiated ordering Hanji yarns in Ssang-Young to
produce its branded sportswear T-shirt lines in 2015 (The North Face 2015).
Additional orders of Hanji yarn will be made by The North Face depending on
market response (The North Face 2015). The North Face Korea is operated by
Youngone Outdoor, a sister company to Youngone Corporation, which is a multinational manufacturer of outdoor and athletic clothing, functional textiles, foot-
wear, and gear (The North Face 2015). Youngone has strong buying power for
raw materials because it makes US $1.6 billion in annual sales by manufactur-
ing products for outdoor sportswear companies (Younone.co.kr N.D.). The North
Face's order of Hanji yarn shows meaningful economic support for local textile
mills in the Jeolla province of Korea. It also triggers the creation of additional
business opportunities with other large corporations such as Kolon Sports and
Blackyak (Northern Jeolla Province Daily 2014).

Fig. 10 The North Face Korea's Hanji yarn T-shirts (The North Face Korea 2015) and ECOTEC (Daegu Daily 2012)

Youngone's Chairman and CEO Kihak Sung currently serves as a chair of the Korea Federation of Textile Industries (KOFOTI), an organization supporting textile and apparel companies founded in 1967 (Kofoti N.D.). Chairman Kihak Sung has made significant contributions to the further consumption of green materials in the fashion industry. He supports local mills, which manufacture eco-friendly materials, including Hanji yarn, by expanding distribution channels, increasing export opportunities, and enhancing R&D capabilities. KOFOTI hosts symposiums to establish the future growth strategies of the green fashion businesses under Chairman Kihak Sung's guidance (Northern Jeolla Province Daily 2014). KOFOTI established ECOTEC in 2012, a Korean eco-label to certify green quality and management of textile products and companies (KOFOTI 2012). See Fig. 10.

Hyundai Dymos is a subsidiary of Hyundai Motors, manufacturing car interior coverings and accessories; they presented eco-friendly conceptual car-seat cover lines made with Hanji fabric in 2014 (Hyundai Dymos 2015). This was the first case of the adoption of sustainable traditional fabrics in the automobile industry in Korea. The Hanji fabric car-seats were designed for Hyundai Motor's luxury sedan, Genesis 2015 (Naver 2014). There were two seat-coverings and nine backboard garnishes designed in collaboration with artisans at ZIIN Company who specialize in craft design and development using Korean traditional materials, including Hanji (ZIIN 2015). Eight garnish designs were obtained for Korean design patents in 2015 (Hyundai Dymos 2015). According to Hyundai Dymos (2015), naturally dyed Hanji fabrics that embodied the figure of traditional Korean Bojagi (a patch worked wrapping cloth) were applied to the upper part of the seat-coverings to minimize the surface friction of the seat. Korean traditional papercraft

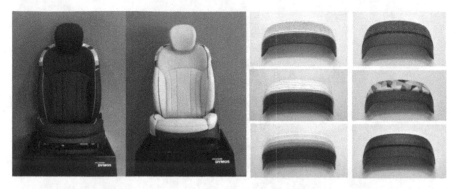

Seat covers (upper shoulder area) and Backboard garnish covers

Seat and garnish cover making processes

Fig. 11 Hanji fabric seat-coverings developed by Hyundai Dymos and ZIIN (Hyundai Dymos 2015; ZIIN 2015)

techniques, such as Ji-Seung, Ji-Ho, and Jeon-Ji, were performed to dress the backboard garnishes (ZIIN 2015). Developed seat-coverings and backboard garnishes were not able to be commercialized because of limitations associated with mass production of elaborate Hanji fabric handcrafts for seat-coverings (Hyundai Dymos 2015). However, Hyundai Dymos plans to continue supporting the integration of natural materials into their car interior products in the future (Hyundai Dymos 2015). See Fig. 11.

5.4 Governmental Support

The Korean government has been aggressively promoting Hanji yarn and eco-friendly materials to businesses in Korea (Korea Fashion and Textile News 2015a, b).

There are governmental funds and grants available for natural material manufacturers and research institutions in Korea. For instance, the Northern Jeolla provincial government supported seven local Hanji yarn manufacturers in the R&D of Hanji yarn related products for a total of US $0.3 million in 2015 (Financial Services Commission of Korea 2015). The city of Jeonju has been hosting an annual Hanji Culture Festival, which has presented Hanji textiles and apparel products and paper handicrafts for 20 years (Naver 2014). The Korean government declared green growth as the strategic priority of the Korean textile industry (Ministry of Government Legislation 2010).

6 Future of Hanji Yarn in the Apparel Industry

The Hanji yarn business in Korea has grown rapidly, reaching US $61.82 million in global Hanji textile product sales in 2013, and it is expected to increase to approximately US $436.36 million by 2018 (Korea Institute for Knit Industry 2015). Thus, Hanji yarn has a huge potential to generate economic benefits and new employment opportunities for the local provinces in Korea continuously. Korean fabric mills' capability to produce Hanji yarn is forecasted to increase from 550 to 1,400 tons per year (Korea Institute for Knit Industry 2015). Therefore, local Hanji yarn manufacturers may obtain more business opportunities with large apparel companies to procure Hanji yarn in high volume. Hanji yarn textiles' distinctive contributions to the global fashion scene are being presented as a trendy fashion icon in denim and sportswear lines (The North Face 2015; Troa 2015). Hanji yarn products are distributed to a wide range of market segments from high-end to mass merchants globally (Troa 2015).

The lifestyle of health and sustainability (LOHAS) directs careful and respectful consumption behavior towards healthy and sustainable directions (Emerich 2011). LOHAS has become a popular culture in the fashion industry today. LOHAS product market size was approximately US $14.9 billion in 2008 and is growing 2.7 % annually (Oh et al. 2010). In a study conducted by the Natural Marketing Institute (2010), 90 % of American consumers were motivated to buy green products in general and 35 % of them responded their willingness to pay a premium price for green products in the LOHAS consumer survey (Emerich 2011). LOHAS encourages eco-design and contributes greatly to the growth of green textiles such as Hanji fabric in the fashion industry (Emerich 2011).

Currently, many global apparel brands and retailers such as the North Face and Kolon Sports adopt Hanji yarn to make products such as functional sportswear, premium fashion apparel, and home textiles, and their order quantities have been constantly increasing (Korea Fashion and Textile News 2015a, b). Manufacturers and research institutions' continuous efforts in Hanji yarn development will identify the broader range of Hanji yarn usage. Integration of advanced technologies such as application of titania nanorods or silver dust to Hanji yarn suggests new market opportunities to companies specializing in performance textiles or sportswear.

There are some challenges to the growth of the Hanji yarn business. Versatile application of Hanji yarn in the apparel industry is still limited due to weaker tensile strength as a paper-based yarn, despite the constantly developing manufacturing techniques (Korea Institute for Knit Industry 2015). Many Hanji yarn manufacturers currently struggle to match their production capacities to the demands of buyers, except very few companies who are relatively larger with more advanced production techniques and infrastructures, such as Ssang-Young and Oh-Sung (MBC 2014). The Korean government and KOFOTI constantly support local Hanji manufacturers to enhance their production capacities, but there is room to grow the Hanji yarn business in a global marketplace.

7 Conclusion

The present case study about the Hanji yarn business of Korea includes the following: the advanced manufacturing processes, advantages, and the current business state of Hanji yarn in the fashion industry. Manufacturers of Hanji yarn have advanced manufacturing processes by selecting Hanji yarns of proper widths and weights, adding water-supply methods, and differentiating knitting and weaving techniques (Oh et al. 2010; Park and Joo 2012; Park and Lee 2013). Six major advantages of using Hanji yarn have been identified for the environment (e.g., eco-friendliness, fast growth of the mulberry tree), the consumers (e.g., antimicrobial activities, wearing comfort, promotion of health, textile aesthetics), and the apparel industry (e.g., economic benefits to local fabric mills) (Gil et al. 2010; Kew Royal Botanical Gardens N.D.; Korea Institute for Knit Industry 2015; Park and Joo 2012; Xu et al. 2011). The textile and apparel businesses related to Hanji yarn have grown dramatically over the last five years and are expected to continue growing in the future (MBC 2014; Naver 2014). The recent success of Hanji yarn in the textile industry has helped rejuvenate local economies in Korea by generating new opportunities for business and employment (Korea E-daily News 2013). The Korean government and various private organizations avidly support the usage of Hanji yarn and other eco-friendly materials in the fashion industry (Financial Services Commission of Korea 2015). Many global textile manufacturers and apparel companies have invested their efforts in facilitating the adoption of Hanji yarn in their products (Isae 2015; Ssang-Young 2015; The North Face Korea 2015; Troa 2015).

Hanji yarn and its related businesses contribute to the development of sustainable production and consumption of products in the fashion industry. This case study addresses how natural materials can contribute to rejuvenate nature and the sustainable fashion industry of Korea. Green design and management should be the priorities in textile and apparel companies. The sustainable business practice of utilizing natural fibers such as Hanji yarn is important for not only the benefits to the apparel companies, but also for consumers and for academics preparing students to use natural materials in the fashion industry.

Future research should be conducted regarding the potential growth of global business opportunities within Hanji yarn products. Standardized assessment methods should be examined to evaluate the green design and management of companies involved in Hanji yarn manufacturing. Topics related to the integration of advanced technologies into Hanji yarns will be important because they could potentially add additional values to the natural materials. There were limitations in identifying the current business state of Hanji textile manufacturers and retailers in international countries such as the United States and China. If there was accessible information for Hanji textile companies overseas, it would have been helpful to determine more accurate facts regarding the Hanji yarn businesses from a global perspective.

Appendix

Table 3

Table 3 Hanji yarn related company listing

Category	Name	Contact	Phone	Location	Website
Manufacturer	Oh-Sung	CEO/Lee, Junyup	82-63-212-0661	Jeonjoo, Korea	N/A
	Ssang-Young	CEO/Kim, Kang-Hoon	82-63-830-5114	Iksan, Korea	www.Lohashanji.com
	Sung-Sil	CEO/Chung, Tae-Doo	82-63-272-0762	Jeonjoo, Korea	www.Sung-Sil.com
Research institution	Korea Institute for Knit Industry	General Director/Baek, Chul-Kyu	82-63-830-3500	Iksan, Korea	www.knitcenter.re.kr
Private organization	Korea Federation of Textile Industry	Chairman/ Sung, Ki-Hak	82-2-528-4058	Seoul, Korea	www.kofoti.or.kr
Apparel company	Isae FNC	CEO/Chung, Kyung-A	82-2-763-6818	Seoul, Korea	www.isae.co.kr
	Troa	CEO/Song, Han	201-961-2211	New York, USA	www.troaco.com
	Youngone Corporation	Chairman/ Sung, Kihak	82-2-390-6114	Seoul, Korea	www.youngone.co.kr
Automobile company	Hyundai Dymos	CEO/Yeo, Seung-Dong	82-41-661-7061	Seo-San, Korea	www.hyundai-dymos.com
Artcraft design company	ZIIN	CEO/Han, Sora	82-2-766-5201	Seoul, Korea	www.Ziin.co

References

Abreu M-C-S (2015) Perspectives, drivers, and a roadmap for corporate social responsibility in the textile and clothing industry. Text Sci Clothing Technol 1–21

Adams C-A, Frost G-R (2008) Integrating sustainability reporting into management practices. Acc Forum 32(4):288–302

Aliagha G-U, Cin N-Y (2013) Perceptions of Malaysian office workers on the adoption of the Japanese Cool-Biz concept of energy conservation. J Asian Afr Stud 48(4):427–446

Choi J-I, Chung Y-J, Kang D-I, Lee K-S, Lee J-W (2012) Effect of radiation on disinfection and mechanical properties of Korean traditional paper, Hanji. Radiat Phys Chem 81(8):1051–1054

Daegu Daily (2012) KOFOTI established ECOTEC. http://daeguilbo.blog.me/40167240292. Accessed 15 Jul 2015

Dymos Hyundai (2015) Hanji car-seat covering project. Kim, B.R, Seoul, Korea

E-daily News (2013) Ssang-Young's CEO, Kang-Hoon Kim. http://www.edaily.co.kr/news/New sRead.edy?SCD=JC61&newsid=01128326602875832&DCD=A00306&OutLnkChk=Y. Accessed Aug 7 2015

Emerich M-M (2011) The Gospel of sustainability: media, market, and LOHAS. Illinois, Champaign, p 232

Fetcher A (2012) Patagonia continues to lead base layer market with Fall'13 collection, 100 % Blue Sign ® system approved fabric. http://www.patagoniaworks.com/Press/2014/6/24/patagonia-continues-to-lead-baselayer-market-with-fall-13-collection-100-bluesign-approved-fabric

Financial Services Commission of Korea (2015) Government funding for sustainable manufacturing industry using Hanji yarn. http://www.smefn.or.kr/articleDetail.do?main._sno=22279. Accessed 15 Jul 2015

Gardetti M, Muthu S (2015) Sustainable apparel? Is the innovation in the business model? - The case of IOU project. Text Clothing Sustain 1(1):1–9

Gil M-S, Jeong U-Y, Son H-J (2010) Pulping automation and manufacturing technology of environmental friendly mulberry fiber. J Korean Fiber Soc 14(2):78–85

Hankook Daily (2009) Three companies in Northern Jeolla province conquered the down-Economy. http://economy.hankooki.com/lpage/society/200901/e2009012718242893840.htm. Accessed Aug 7 2015

Hansmann K-W, Claudia K (eds) (2001) Environmental management policies. Greenleaf, Sheffield, pp 192–204

International Organization for Standardization (2004) ISO 14001:2004 Environmental management systems—Requirements with guidance for use. http://www.iso.org/iso/home/store/catalogue_tc/catalogue_detail.htm?csnumber=31807. Accessed 15 Jul 2015

Isae.co.kr (N.D.) Process. http://www.isae.co.kr. Accessed 20 Jul 2015Isae (2015) Hanji fabric and natural dyeing. Seoul, Korea: Lee, M.A

Jang Y-S, Amna T, Hassan M-S, Kim H-C, Kim J-H, Baik S-H, Khil M-S (2015) Nanotitania/mulberry fibers as novel textile with anti-yellowing and intrinsic antimicrobial properties. Ceram Int 41:6274–6280

Japan Today (2014) Warm-Biz campaign kicks off across Japan. http://www.japantoday.com/national/view/warm-biz-campaign-kicks-off-across-japan. Accessed 15 Jul 2015

Jeong M-J, Kang K-Y, Bacher M-K, Kim H-J, Jo B-M, Potthast A (2014) Deterioration of ancient cellulose paper, Hanji: evaluation of paper permanence. Cellulose 21(6):4621–4632

Ju J-A, Shim J-Y, Kim H-C (2013) Methods to improve infant clothing made with Hanji yarn—investigating the image of Hanji and Hanji yarn infant clothing. Res J Costume Culture Korea 21(1):57–65

KEW Royal Botanical Gardens (N.D.) Broussonetia papyrifera (paper mulberry). http://www.kew.org/science-conservation/plants-fungi/broussonetia-papyrifera-paper-mulberry. Accessed 11 Aug 2015

Khan M-R, Islam M (2015) Materials and manufacturing environmental sustainability evaluation of apparel product: knitted T-shirt case study. Text Clothing Sustain 1(1):1–12

Kim D-K (2006) The natural environment control system of Korean traditional architecture: comparison with Korean contemporary architecture. Build Environ 41(12):1905–1912

Korea Fashion and Textile News (2015a) Korea Institute for Knit Industry launches Warm-BIz workwear using Hanji yarn this winder. http://www.ktnews.com/sub/view.php?cd_news=94586. Accessed 15 Jul 2015

Korea Fashion and Textile News (2015b) A new Hanji yarn developed for Outdoor sportswear and casual apparels. http://www.ktnews.com/sub/view.php?cd_news=94514. Accessed 11 Aug 2015

Korea Federation of Textile Industries (2012) ECOTEC: Korea Eco Textile Certification. http://www.kofoti.or.kr/OpBoard/View.asp?Code=KNM&Uid=936. Accessed 15 Jul 2015

Korea Institute for Knit Industry (2015) Warm-Biz project. Lee, S.J, Seoul, Korea

Lewis H, Gertsakis J, Grant T, Morelli N, Sweatman A (2001) Design environment: a global guide to designing greener goods. Sheffiled, Greenleaf, p 16

MBC (2014) Textile business is rejuvenated with incorporation of technologies. http://imnews.im bc.com/replay/2014/nwtoday/article/3521072_13495.html. Accessed 15 Jul 2015

Ministry of Government Legislation (2010) Framework act and its presidential decree on low carbon, green growth in Korea. http://m.moleg.go.kr/english/korLawEng?pstSeq—57719& brdSeq=33&pageIndex=10. Accessed 15 Jul 2015

Natural Marketing Institute (2010) LOHAS around the world. LOHAS Journal Spring 11–15

Naver (2014) Hyundai Dymos participated in 2014 Seoul Hanji Cultural Festival. http://blog.nav er.com/seoulhanji/220136014530. Accessed 15 Jul 2015

Northen Jeolla Province Daily (2014) Symposium of KOFOTI to expand distribution channels of textile mills in Jeolla. http://www.domin.co.kr/news/articleView.html?idxno=1046602 Accessed 15 Jul 2015

Oecotextile (2013) Climate change and the textile industry. https://oecotextiles.wordpress.com/? s=apparel+industry Accessed 15 Jul 2015

Oceotextiles (N.D.) Textile industry poses environmental hazards. http://www.oecotextiles. com/PDF/textile_industry_hazards.pdf. Accessed 18 Jul 2015

Oh D-H, Na H-J, Jeon Y-H, Jeong W-Y, An S-K (2010) A Study on the physical propertiesof woven fabric with Hanji Yarn. Text Sci Eng Korea 47(4):272–278

Park T-Y, Joo C-W (2012) Evaluation of tensile property and surface structure of Hanji paper yarn with different twist numbers. Text Sci Eng Korea 49(3):147–152

Park T-Y, Lee S-G (2013) A study on coarse Hanji yarn manufacturing and properties of the Hanji fabric. Fibers Polym Korea 14(2):311–315

Takasaki M, Ogura R, Morikawa H, Chino S, Tsuiki H (2011) Preparation and properties of paper yarn from mulberry. Adv Mater Res 175–176:575–579

Pigosso D-C-A, Rozenfeld H, McAloone T-C (2013) Ecodesign maturity model: a management framework to support ecodesign implementation into manufacturing companies. J Clean Prod 59:160–173

Roy Choudhury A-K (2015) Development of eco-labels for sustainable textiles. Roadmap to Sustainable Textiles and Clothing, Text Sci Clothing Technol 137–173

Saito K, Linquist B, Keobualapha B, Shiraiwa T, Horie T (2009) Broussonetia papyrifera (paper mulberry): its growth, yield and potential as a fallow crop in slash-and-burn upland rice system of northern Laos. 76(3): 525–532

Ssang-Young (2015) The Hanji business in Korea. N.A, Ilsan, Korea

Sivaramakrishnan C-N (2009) Pollution in textile industry. Colourage 16(2):66–68

Steiner J, Steiner G (2009) Business, government, and society: managerial perspective. Text and case. 12th edn. McGraw-Hill Irwin, New York

Sustainable Apparel Coalition. (N.A.) The Coalition. http://apparelcoalition.org/the-coalition/

The North Face Korea (2015) Hanji T-shirt. Jeong, W.T, Seoul, Korea

Troa (2015) Hanji denim lines and natural dyeing. New York, NY: Song, H.

Troaco.com (N.D.) Troa: Hanji to denim. https://vimeo.com/86148250. Accessed 15 Jul 2015

UNESCO (N.D.) Memory of the world. http://www.unesco.org/new/en/communication-and-information/flagship-project-activities/memory-of-the-world/register/access-by-region-and-country/asia-and-the-pacific/republic-of-korea. Accessed 15 Jul 2015

Xu Y, Wong M, Yang J, Ye Z, Jiang P, Zheng S (2011) Dynamics of carbon accumulation during the fast growth period of bamboo plant. Bot Rev 77(3):287–295

Yonhap News (2010) Sportswear made with Hanji yarn. http://news.naver.com/main/read.nhn?mode=LSD&mid=sec&sid1=102&oid=001&aid=0003100560. Accessed 7 Aug 2015

Youngone.co.kr (N.D.) Youngone's present. http://www.youngone.co.kr. Accessed 7 Aug 2015

Yum H, Singer B-W, Bacon A (2009) Coniferous wood pulp in traditional Korean paper between the 15th and 18th centuries. Archaeometry 51(3):467–479

Zhu O, Sarkis J, Geng Y (2005) Green supply chain management in China: pressures, practices and performance. Int J Oper Prod Manage 25(5):449–468

ZIIN (2015) Hanji car-seat covering project. Han, S.R, Seoul, Korea

Sustainable Production Processes in Textile Dyeing

L. Ammayappan, Seiko Jose and A. Arputha Raj

Abstract Value addition for textiles is an important process and it is required for all products including yarn, fabric, garment, fashion apparel, floor covering, and the majority of technical textiles. Value addition may be either an additive or subtractive process. In the subtractive process part of the fiber components may be removed by some physical or chemical process because textile fibers have natural as well as added impurities during production. In the additive process, either color or functional chemicals may be added to improve aesthetic as well as functional properties. For each textile processing, the processor used enormous amounts of chemicals and water in order to attain the desired result. After processing the residual processed chemicals and waters are treated with effluent treatment and discharged into the mainstream. The amount of residual unfixed dyes, metal compounds, formaldehyde-based dye-fixing agents, hydrocarbon-based softeners, and all types of dye-bath auxiliaries as well as their degradation nature against the environment are the deciding factors for its sustainability. Technologies have been developed for the past five decades starting from fiber to finished product in order to reduce the effluent load, energy, processing cost, and manpower as well as increase the process efficiency and reproducibility. There are many factors influencing the overall efficiency or value addition of a textile product, which play an important role in its sustainability. This chapter deals with the basic theory of dyeing processes, factors influencing their performance, potential pollutants, sustainable technologies developed thus far, and future perspectives in dyeing.

Keywords Dyeing · Pollution · Natural fiber · Pretreatment

L. Ammayappan (✉) · S. Jose
ICAR-National Institute of Research on Jute and Allied Fiber Technology,
12 Regent Park, Kolkata 700040, West Bengal, India
e-mail: lammayappan@yahoo.co.in

A. Arputha Raj
ICAR-Central Institute for Research on Cotton Technology, Adenwala Road,
Matunga (East), Mumbai 400 019, India

© Springer Science+Business Media Singapore 2016
S.S. Muthu and M.A. Gardetti (eds.), *Green Fashion*,
Environmental Footprints and Eco-design of Products and Processes,
DOI 10.1007/978-981-10-0111-6_8

1 Introduction

Humans need food for survival; dwellings for protection against nature such as sunlight, rain, and storms; as well as privacy and clothing for civilization with protection of the body against the environment. In older days, the fashion sector generally designed clothing as per requirements and desire; hence they wanted to differentiate from others. After the invention of textile fiber followed by hand spinning/hand loom weaving, artisans introduced coloration on clothing. They strongly believed that the color perceived by humans from nature could change the mood as well as the personality. Soon they identified different natural dyes from various parts of plants and some insects, applied to fabrics by different methods and termed dyeing or coloration. Dyers developed their own dyeing methodology from their experience and traditional knowledge. Successful dyeing technology based on most of the natural dyes were kept secret in many countries for many years, because such dyeing technology was believed to be a prestige art in ancient history and executed only by skilled persons.

Depending upon the durability of color on fiber and aesthetic appeal of the clothing, every dyeing process is distinguished by the end-user; for example, empire clothing should be dyed with purple/crimson red for a royal look; saint clothing is dyed with orange color for an indication of their devotion to God; devotee clothing is dyed with golden yellow color for their positive promise to religion. Natural dye coloration has limitations due to the limited number of colors, medium fastness properties, low yield, and nonreproducibility. After invention of synthetic dyes, dyers preferred to go with coloration based on synthetic instead of natural dyes due to the wide range of colors, good fastness properties, ease of application, and reproducible shades. After modernization of the dyeing industry, processors set up their dyeing industry as per their end-products and the production capacity. There was no stringent restriction on discharge of unfixed dyes/chemicals from the textile-processing industry as well as the amount of eco-friendly dyes/finishing chemicals present on clothing before the 1990s. During the 1990s, there was an awareness of compulsory effluent treatment for wastewater discharged from all processing industries and introduction of "carcinogenic chemical free" clothing. R&D laboratories have identified carcinogenic dyes/chemicals; subsequently the European Union (EU) banned those dyes/chemicals. Testing authorities have introduced various standards for eco-friendly/green products. Soon it became mandatory to test the finished clothing for eco-friendliness. Today for an exportable textile material to a developed country, the supplier should provide suitable certification for the product such as Oeko-Tex, Blue Sign, iVN, KRAV, JOCA, REACH, ZDHC, GOTS, and Eco label.

Some of the synthetic dyestuffs are considered harmful to humans and the environment. Awareness of the harmfulness of synthetic dyestuffs led to the development of eco-friendly dyeing methodologies, green dyes, and energy-saving machinery/process. Dyeing is one of the large-scale processes and so scientists are focused on each and every raw material/process for their sustainability. Over the past years, dyers follow only sustainable dyeing technology; however, R&D works also have been used to fine tune the existing methodology to narrow their target and to enhance the quality of the product. It is the right time to overview the types of dyes and their application on fibers, existing dyeing practices, dyeing mechanisms, environmental issues due to dyeing, and alternative potential dyeing technology for sustainability.

This chapter explores the suitable synthetic dyes for dyeing, factors responsible for improving dyeability, potential areas in processing methodology, material selection, machinery development, and management for sustainable dyeing technology with special reference to natural fiber based clothing.

2 Natural Fiber

Natural fiber is derived from a biopolymeric material and described by the flexibility, fineness, and specifically high length to transverse dimension ratio. Natural fibers have been used for the development of fashion textiles since civilization began. Natural fibers are biopolymers of long chain molecules, formed by the polymerization of cellulose or protein monomer. The fixation of dyes inside the fiber network is mainly governed by the morphology and chemical composition of the fiber. For example nonionic disperse dyes have good affinity towards polyester fibers at high temperature, whereas anionic acid dyes have good fixation on protein fibers. Good understanding of the morphology and characteristics of textile fibers could suggest the proper selection of dyes with good fastness properties, color depth, and suitable dyeing process in economic and eco-friendly points of view.

Each natural fiber is differentiated from others in terms of physical, chemical, and mechanical properties. Depending on the comfort and ease of processing, their demand by the textile market may be varied. Each natural fiber has its own dyeing methodology; however, cellulosic fibers or protein fibers have a common dyeing protocol. Classification of natural fibers based on their chemical constitution is given in Table 1 (Mussig and Stevens 2010).

3 Dyeing Process

Humans have always been fascinated with color, because undyed clothing does not normally attract attention. To make different designs and aesthetic appeal in clothing, experts introduced the dyeing process. Dyeing is a textile wet processing in which color is incorporated into fibrous products in different forms such as loose fiber, yarn, fabric, and nonwoven in a suitable dyeing machine. Before dyeing, preparatory processes are required to remove natural as well as added impurities present in the fiber. The sequence of preparatory processes for each product

Table 1 Classification of natural fibers based on sources

Cellulosic fibers				Protein fibers			
Seed	Bast	Leaf	Fruit	Animal hair		Fur	Silk
				Sheep and goat	Camel and camelids		
• Cotton	• Jute	• Sisal	• Coir	• Wool	• Camel	• Horse	• Silk
• Milkweed	• Flax	• Henequen	• Sugar palm		• Alpaca	• Musk Ox	• Spider silk
• Kapok	• Hemp	• Abaca	• Palm	• Mohair	• Vicuna	• Angora rabbit	
• Cattail	• Kenaf	• Pineapple		• Pashmina	• Guanaco	• Hare	
	• Mesta	• Date Palm		• Cashmere	• Llama	• Bison	
	• Ramie	• Manila hemp		• Cashgora	• Vicuna	• Yak	
	• Banana	• New Zealand flax		•Pygora	• Antelope		
	• Bamboo						
	• Sunnhemp						

may be different, depending on the characteristics of the dye and the form of the substrate (Trotman 1984).

3.1 Preparatory Processes

The main objective of the preparatory process is to make the textile substrate ready for subsequent processing. It might be either an additive process by giving additional functionality to the fiber or a subtractive process by removing the unnecessary impurities (Madaras et al. 1993; Trotman 1984).

3.1.1 Desizing

Desizing removes the added sizing material either by acid- or enzyme-assisted hydrolysis. Presence of any sizing materials such as starch, polyvinyl alcohol, and guar gum, could hinder the penetration of dyes and chemicals inside the fiber. Desizing is carried out by hydrolysing the sizing material by any one of the following chemicals: amylase enzyme, sodium persulphate, oxidizing agents, and hydrochloric acid. Among them, enzyme-based desizing is preferred for its sustainability (Cavaco-Paulo and Gubitz 2003).

3.1.2 Scouring

Scouring removes the adhered as well as added fatty matter present in the textile material by a saponification process. Natural fibers have impurities including pectin, fat, hemicelluloses, oil, minerals, natural coloring matter, and wax. These impurities are mainly acidic in nature and can be easily removed by a hot alkaline condition. Recently lipase-based scouring and combined scouring and bleaching have been preferred in terms of economy and sustainability (Ammayappan ct al. 2003)

3.1.3 Bleaching

Bleaching removes the natural coloring matter present in the natural fibers by either reducing or oxidizing agents. Processors prefer hydrogen-peroxide–based bleaching for its durability and sustainability. For dark color shades a combined scouring and bleaching process is preferred in order to save water and energy (Ammayappan et al. 2003).

3.1.4 Carbonizing

Carbonizing removes vegetable-based impurities such as burrs, seeds, or dust adhered to the greasy wool fleece by acid backing followed by neutralization. Sulphuric acid converts the cellulose biomass into dehydrocellulose followed by charred mass and it can be removed by crushing and beating. This process is generally preferred for light color shades (Ammayappan 2009).

3.1.5 Degumming

Degumming removes the gummy substance (20–25 % sericin) present in the silk cocoon by the hot alkaline solution. Degumming improves the luster, color, hand, and texture of the silk materials.

3.1.6 Mercerization

Mercerization is used on cellulosic textiles with 250 g/L of NaOH for 180 s at 5 °C in order to improve its reactivity and luster. Mercerization partially converts the cellulose molecule into soda cellulose and that can improve the dyeing as well as fixation rate of dyes in light to dark shades. Reduction in concentration of sodium hydroxide to attain the mercerized effect is the focus of this process for its sustainability (Marsh 1941)

4 Dyes

Dye is an organic molecule, used to impart color to the textile fiber. Each dye molecule used for natural fiber has the following three groups: chromophore, auxochrome, and solubilizing. Chromophore provides a distinct color to the dye molecule due to absorption of certain wavelengths of visible light and subsequent transmission/reflection of other regions. An auxochrome is a functional group attached to the chromophore and it modifies the wavelength and intensity of absorption of light, so that it intensifies the color produced by the chromophore. The solubilizing group makes the dye molecule water soluble. For example: Azo group ($-N=N-$) is a chromophore; $-OH$, $-NH_2$, $-OCH_3$ is an auxochrome; $-SO_3Na$ is a solubilizing group. Dyes are classified based on their mode of application and the different types of dyes with their sustainability have been discussed in the following (Venkataraman 1971; Broadbent 2001; Bird and Boston 1975).

4.1 Acid Dyes

Acid dyes are sodium salts of sulphonic acid or carboxylic acids ($R-SO_3Na/R-COONa$) and applied in acidic pH on wool and silk fibers. The majority of azo-based acid dyes are banned due to their carcinogenicity. Today, these groups of dyes have been replaced with metal complex dyes (Lewis 1992; Ammayappan 2003).

4.2 Basic Dyes

Basic dyes are halides of tertiary aromatic ammonium compounds that can bond with the functional group of textile fiber such as $-COO^-$ or $-O^-$ during dyeing. They are mainly applied to protein fibers, acrylic fiber, and tanned cellulosic fibers especially for bright shades. Basic dyes did not have good washing as well as light fastness properties; they can be improved by after-treatments with tannic acids or direct dyes. Silk fashion textiles prefer selective basic dyes that have eco-sustainability for their brilliant shade (Ingamells 1993)

4.3 Direct Dyes

Direct dyes have very good affinity towards all kinds of natural fibers in the absence of any additives. Direct dye-dyed materials show moderate washing and light fastness properties due to weak hydrogen bonding with the functional groups of fiber. However, fastness properties can be improved by suitable cationic fixing agents. Azo- and benzidine-based direct dye consumption in the centralized sector is reduced due to its carcinogenicity, however, cottage industries are still using them due to their low cost (Cegarra 1998; Shore 1995).

4.4 Mordant Dyes

Acid dyes have moderate fastness properties due to their poor aggregation inside the fiber matrix. To improve the aggregation of anionic dyes, the mordanting process is introduced in which a metallic compound is applied to the substrate. However, mordant dyes are banned due to the carcinogenic nature of the chromium metals as well as lengthy processing (Angelini et al. 1997; Mendez et al. 2004; Cho 1999).

4.5 Metal Complex Dyes

Metal complex dyes are premetalized dyes; that is, one or two dye molecules are coordinated with a metal ion during synthesis. Metal complex dyes are of two types, 1:1 and 1:2 metal complex dyes, in which one metal cation is bonded with one and two dye molecules, respectively. These dyes are mainly applied to woolen and silk textiles for dull shades and good aggregation, so that this class of dyestuff has good sustainability (Lewis 1992; Ammayappan 2003).

4.6 Reactive Dyes

Reactive dye has a chromophore, one or two reactive auxophores, and a solubilizing group. This dye forms covalent bonding with functional groups of the fiber through either substitution or addition reaction and is mainly applied to cellulosic fibers in the presence of an electrolyte for exhaustion and an alkali for fixation. Recently bifunctional reactive dyes have been introduced in which both triazine and sulphato ethyl sulphone reactive groups are present in a single dye molecule in order to enhance the dye fixation and fastness properties at a moderate dyeing temperature. Low salt/alkali fixation reactive dye, low MLR dyeing machine, and excellent bleaching fastness are the important sustainable factors (Broadbent 2001).

4.7 Vat Dyes

Vat dyes are anthraquinone or indigo based insoluble dyes and are traditionally used to dye cellulosic textiles in a vat. Solubilization and stabilization of sodium salt of leuco vat dye is the crucial factor and it consumes more chemicals and energy than other parameters. R&D interventions are mainly focused on reducing the consumption of the reducing agent in order to sustain this dyeing process (Broadbent 2001).

4.8 Sulphur Dyes

Sulphur dyes are amino/nitro aromatic compounds with $-S=S-$ linkages and most popular for black and brown shades on cellulosic textiles. Men's fashion wear for wild black and brownish shades are developed mainly from sulphur dyes. Recently, sulphur dyes have been slowly phased out due to the polluting nature of the dye-bath effluent. Dyeing methodology using glucose for the substitution of sulfide-reducing agents and reducing dye by electrochemical processes are the promising sustainable processes (Shore 1995; Broadbent 2001).

4.9 Solubilized Vat Dyes

Solubilized vat dyes are sodium salt sulphuric esters of the leuco vat acid and applied to cellulosic textiles directly in a neutral or slightly alkaline condition. They are famous for durable and elegant pale shades on cellulosic textiles. This class of dyes was never widespread and has declined due to their high cost (Broadbent 2001).

5 Influencing Factors for Dyeing Process

The dye molecule is a complex organic compound with one chromophore and one or more auxophores for its fixation with textile fibers. In aqueous solution, it exists either as an individual or aggregate form and the state depends on the ionic nature and molecular weight of the dye molecule. Dyeing involves the following three important steps (Peters and Vickerstaff 1948; Broadbent 2001; Ingamells 1993).

- Migration of the dye molecules from the dye bath towards the surface of the fiber and subsequent adsorption on the fiber surface
- Diffusion of the dye molecule from the surface to the inside matrix of the fiber due to the concentration gradient of the dye molecule
- Fixation or bonding of the diffused dye molecules with the functional groups of fiber polymer through different chemical bondings followed by orientation and aggregation (Johnson 1989)

Among the three steps, the diffusion process is the rate-determining step. The diffusion behavior of a dye molecule differs in each fiber and it helps to select the suitable dye as per the dyer's requirements such as the cost of dyeing, durability, and end use. The diffusion of the dye molecule inside the fiber matrix depends on the dye molecular size, dye molecular shape, ionic nature of the dye, diffusion coefficient, dyeing adsorption profile, substrate type, substrate preparation, and water quality (Madaras et al. 1993). The dyeing process depends on many factors in order to get uniform and solid shades with excellent fastness properties (Johnson 1989; Madaras 1993; Ingamells 1993). These factors are discussed in the following chapter in terms of their role in the sustainability of the dyeing process.

5.1 Pretreatment

During the dyeing process, if the adsorption of the dye molecule on the surface of the fiber surface is good, then there will be uniform and solid dyeing. So pretreatment is required to improve the wettability as well as functionality of the fiber that can lead to uniform dyeing.

5.2 Dye Concentration

Dyeing can be carried from pastel shades to dark shades in order to match the requirements of various customers. For pale shades, the dyeing process requires special attention to attain a uniform shade whereas for dark shades it requires additional time, dyeing auxiliaries, and after-treatments in order to improve its fastness properties.

5.3 Dyeing Condition

The diffusion of dye molecules from the surface to the inside of the fiber matrix is the rate-determining step. Dyeing time, dyeing temperature, and material-to-liquor ratio play important roles to enhance the dyeing rate. Dyeing time depends on the affinity of the dye towards the fiber substrate as well as the method of dyeing. High molecular weight/less affinity dye molecules require a longer duration to diffuse and form bonding with the fiber polymer than low molecular weight/high affinity dye molecules. If the affinity of the dye molecule towards the natural fiber is good, dyeing can be done at low temperature such as cold brand reactive dyes, that is, dyed at 30 °C. However, medium to high temperature (60–95 °C) is required to attain good aggregation of dye molecules and appreciable fastness properties (Bae et al. 1997). The material-to-liquor ratio (MLR) indicates the amount of liquor required for processing with respect to the weight of the substrate. MLR can be selected based on the dyeing machine and dyeing method. Recently a modern jet dyeing machine has been developed with a low MLR dyeing process in order to reduce the cost (Rys and Zollinger 1972).

5.4 Dyeing Auxiliaries

Dye bath auxiliaries including leveling agents, surfactants, acid, or alkali are required to keep the dye molecule active and in individual form. Natural fibers form negative zeta potential when immersed in water. Electrolytes such as $NaCl/Na_2SO_4$ are used to neutralize the zeta potential. Acid/alkali maintain the pH of the dyeing bath in order to improve dye fixation (Noah et al. 1986).

5.5 Compatibility of Dyes

Compatibility between dyes is required for combination shades. For a solid combination shade, both dyes should have uniform diffusion behavior, fixation rate, activation energy, and fastness properties, otherwise the dominant dye will be diffused inside the fiber at a faster rate and the second dye will remain in the bath itself.

5.6 Mechanical Agitation

Mechanical agitation generally enhances the migration of the dye molecule from the dye bath to the surface of the fiber substrate, so that dyeing time can be reduced. Without agitation of the dye bath, agglomeration of dye molecules followed by uneven dyeing may occur. If both the substrate and dyeing liquor are moving, it will ensure a high degree of evenness.

5.7 Quality of Water

Water has a high surface tension and a high dielectric constant and so dyeing is generally carried out in an aqueous medium. Presence of a hardness-creating substance in water can create problems during dyeing by reacting with the dye molecules. A dye house prefers soft water of 50 ppm for quality dyeing.

5.8 Dyeing Methods

The dyeing method plays a critical role in the cost of dyeing and selection of dye for a particular substrate with desirable shade and fastness properties.

5.8.1 Exhaust Dyeing

The exhaust dyeing method imparts color to the textile material in a dyeing bath in the presence of thermal energy, chemical additives, water or solvent, and mechanical action of the dye bath or material. Popular exhaust dyeing machines are soft flow, jigger, and jet dyeing machine. It is a batch dyeing process suitable for small (10 kg) to large (1000 kg) lots. Research work has been carried out to reduce the dyeing temperature and dyeing liquor (MLR) to sustain the dyeing method (Shore 1995; Shamey and Zhao 2014).

5.8.2 Pad → Steam Dyeing

The exhaust dyeing method for certain reactive and vat dyes consumes time, water, and energy. To save energy, the pad → dry → steam process has been introduced in which woven fabric in open width form is padded through 50–300 gpL dyeing solution with 100 % expression followed by steaming with saturated steam at 105 °C for 5–10 min. During steaming, dye molecules on the surface of the fiber migrate towards the swollen fiber structure and are fixed. It is a continuous process and consumes low liquor such as MLR of 1:1 or 1:2. However, the fastness properties are less and the effluent discharge is higher than exhaust dyeing (Madaras et al. 1993).

6 Environmental Impact of the Textile Dyeing Industry

The textile sector plays an important role in the gross domestic product (GDP) of some of the developing countries including India, Pakistan, Bangladesh, and Cambodia. Depending upon the population and life style of the people in a country, the consumption of the different textile products may be varied. If there are more consumers, there will be more requirements for clothing and that leads to the establishment of many textile dyeing industries to meet these requirements. The establishment of dyeing industries is based on government policy on installation of effluent treatment plants. The phenomenal growth of the dyeing industry can bring prosperity to a country by giving more employment; however, it is also responsible for the deterioration of environmental surroundings by discharging effluents. Apart from the air and noise pollution, water pollution poses a big threat to the environment due to the discharge of large amounts of liquid effluents into public sewers, inland surface water bodies, rivers, and irrigation land. The discharge of the effluent can pose a wide range of problems that lead to deterioration of a healthy life for the people. Among textile wet processing starting from preparatory processes to chemical finishing, dyeing is one of the major areas that consumes more water than other processing (Correia et al. 1994; Bartea and Bartea 2008)

Recently the demand for textile clothing is increasing proportional to the human population, and so the consumption of synthetic dyes has increased, It is estimated that over 7×10^5 tons of synthetic dyes are annually produced worldwide for dyeing various textile fibers. After dyeing, nearly 200,000 tons of unfixed dyes are discharged to effluents every year due to the inefficiency of the dyeing process (Forgacs et al. 2004; Dos Santos et al. 2007).

The textile dyeing industry uses more than 8000 chemicals and most of the chemicals are reported to be unhealthy to humans. Presently, water consumption for a dyeing process varies from 30–150 L/kg of cloth, depending on the type of dye and substrate as mentioned in Table 2. Apart from dyeing, water is required for washing the dyed materials to achieve desirable fastness properties. The World Bank estimates that 17–20 % of industrial water pollution comes from the textile dyeing treatment given to fabric. These data give an indication of the effect of conventional dyeing systems on the environment (Vijayaragavan 1999).

It is reported that after treating one ton of cotton fabric, the effluent would have 200–600 ppm BOD, 1000–1600 ppm of total solids, and 30–50 ppm of suspended solids contained in a of 50–160 m^3. For woolen textiles for 1 ton of scoured wool the effluent would have 430–1200 ppm BOD, around 6500 ppm total solids in the of 100–230 m^3 (Robinson et al. 2001). Organized textile sectors treated the textile effluents before discharging into water bodies, whereas cottage industries did not conform to the norms set by the pollution control board. Textile effluents are generally hot, alkaline, strong smelling, and colored. The environmental and ecological effects of textile processes are complex issues. Processing procedures and social and legal constraints vary widely from country to country. It is the time to rethink the existing dyeing system and streamline the sustainable dyeing process in order to preserve the natural resources for our future generations.

Sustainability of a dyeing process can be assessed by its ability to discharge a low amount of unfixed dye, that is, effluent, to the drainage. If the aggregation and

Table 2 Water consumption in the textile industry (Moustafa 2008)

Fiber type/make-up		Mean Water Consumption in L/kg Material
(a) By fiber type	Cotton	50–120
	Wool	75–250
	Other natural fibers	10–100
(b) By make-up	Fabric	100–200
	Hosiery	80–120
	Printing	0–400

Table 3 Fixation degree of different dye classes on textile support (EWA 2005)

Dye class	Fiber type	Fixation degree (%)	Interaction between dye and fiber
Acid	Wool, silk, nylon	80–95	Hydrogen bonding, ionic bonding, van der Waals forces
Basic	Acrylic, silk	95–100	Hydrogen bonding, ionic bonding,
Direct	Cellulose	70–95	Hydrogen bonding, van der Waals forces
Metal complex	Wool	92–98	Hydrogen bonding, ionic bonding, van der Waals forces, hydrophobic bonding
Reactive	Cellulose	50–90	Covalent bonding, hydrogen bonding
Sulphur	Cellulose	60–90	Hydrogen bonding, van der Waals forces
Vat	Cellulose	80–95	Hydrogen bonding, van der Waals forces
Disperse	Polyester, nylon	90–100	Van der Waals forces, hydrophobic bonding
Modified basic	Acrylic	95–98	Ionic bonding, van der Waals forces, hydrophobic bonding

chemical bonding between fiber and dye molecule is good, then there will be less discharge of unfixed dyes. Woolen textile exhausts a higher amount of metal complex dyes than other dyes and thus its sustainability is good (Bertea and Bertea 2008; O'Neill 1999). It is reported that the cotton textile dyeing system discharges more effluent than other fibers as mentioned in Table 3 (EPA 1997). Presently attention is focused towards the right-first-time production techniques that reduce color loads in the effluent by optimization of processes.

7 Sustainable Dyeing

Pure textile material, high fixing dye molecules, low MLR dyeing machine, eco-friendly dyeing method, right-first-time dyeing, skilled manpower, well-planned supply chain, management and stringent policy of the government are the important factors to sustain the dyeing industry in order to meet the global requirements of the fashion industry. Each management is responsible for knowing about the eco-friendly approaches starting from the materials selection to supply of the finished garments to the market (Mahapatra 2011; Bide 2014). In the sustainability of the dyeing industries for the smooth supply of fashion garments, the factors discussed below play a major role.

7.1 Materials

7.1.1 Fiber Substrate

The dyeing method is mainly dependent on the type of fibers: that is, wool fiber cannot be selected for dyeing with vat dyes and cellulosic fiber is not suitable for dyeing with acid dyes. From the sustainability point of view woolen/silk textiles exhaust nearly 95 % of dyes, and cellulosic textiles exhaust nearly 60–85 % of reactive dyes. However, in tropical regional countries customers prefer cool and comfort apparel such as cellulosic textiles, whereas dry cool regional countries prefer warm and hydrophobic apparel like woolen textiles. Cotton fiber is available at an economic price, thus dyers prefer cotton textiles for easy marketing and technologists focus on improving the dye fixation.

7.1.2 Pretreatment

Pretreatments are preferred to improve the dyeing property of a textile substrate. Pretreatment can be classified as physical and chemical pretreatments and it can also improve the appearance, hand, performance, and durability (Ammayappan 2013a). Some pretreatments mainly focus on the surface of the textile substrate and it may be either an additive or subtractive process

Physical Pretreatment

Surface modification of textile fibers in the absence of an aqueous system is termed physical pretreatment. Physical treatment is successful in a pilot-scale study only, due to the huge investment involved and it is less compatible with the conventional processing supply chain. Ultraviolet radiation (UV) treatment, low temperature plasma (LTP) treatment, gamma (δ) radiation, ozone treatment, laser treatment, microwave treatment, e-beam irradiation, and ion implantation are the important physical treatments to improve the dyeability of fibers (Atav 2014).

UV Radiation Treatment

Ultraviolet radiation is an electromagnetic radiation with a wavelength from 400 to 100 nm. UV has the ability to form free radicals and modifies the wool fiber surface to improve its dyeability. It is reported that aromatic amino acid and cystine residues in the wool polymer act as UV-absorbing species and thus absorb UV radiation that lead to oxidize the wool fiber surface in the presence of air. UV-radiated wool fiber has more functional groups than conventional wool fiber and so this photomodification increases the affinity towards anionic dyes at temperatures lower than 90 °C. This technology also improves the production efficiency of wool fabrics especially for heavy shades, especially black and navy blue (Ammayappan 2013b).

Low Temperature Plasma Treatment

Plasma is the fourth state of matter and it is a mixture of electrons, ions, and free radicals. It is produced from an electrical discharge either under vacuum or atmospheric pressure. Plasma has the potential to rupture any polymeric surface physically because it has high activation energy. Plasma treatment in the presence of nonpolymerizing gases such as air, oxygen, and nitrogen etches the outer surface of the fiber up to 30–50 nm depth as well as partial surface oxidation. Physical etching is responsible for the abrasion of the outer layer and surface oxidation introduces new anionic groups, that is, carboxylate groups in the cellulose chains and sulphonate groups in the cuticle cells of wool fiber. This surface modification enhances the dyeability of fiber and is considered an effluent-free and environmentally friendly process (Ammayappan et al. 2012).

Ozone Treatment

Ozone is a powerful oxidizing agent, used to oxidize the textile fiber surface for improving its dyeability through formation of additional functional groups. Ozone is generated through corona discharge, UV light, and electrolysis. Atav and Yurdakul (2011) reported that ozonation on luxury fiber improved the dyeability at lower temperatures by modifying the fiber surfaces.

Gamma Radiation and Others

Gamma radiation denoted γ has electromagnetic radiation of high frequency (above 10^{19} Hz) and energy (above 100 keV). Like UV radiation, gamma radiation, e-beam irradiation, and ion implantation are used to modify the fiber surface, however, due to the high cost, these processes are not common (Beevers and McLaren 1974; Atav 2014).

Chemical Pretreatment

Chemical pretreatment for a textile substrate is mainly based on the chemical nature of dye as well as substrate. If anionic dyes can be selected for dyeing, the textile substrate should be modified to cationic nature by the additive process so that dyeability can be improved as with chitosan treatment or quaternary ammonium compounds on protein fibers (Wei 2009; Schindler and Hauser 2004). There are many chemical treatments reported in the literature, and selective chemical treatments (mainly quaternary ammonium compounds and enzymes) are mainly applicable in terms of a commercial point of view.

Chitosan

Chitosan is a natural biopolymer, chemically called beta-(1, 4)-2-(amino)-2-deoxy-D-glycopyranose. It has an amino group and protonates to NH_3^+ at acidic pH < 6.5 as a cationic polyelectrolyte and it can easily interact with anionic dye molecules. Chitosan-treated wool fiber has more cationic groups than untreated wool fiber and thus a short dyeing time can be reduced. It is also used to improve the dyeability of immature cotton fiber. However, this treatment may impart a stiff feel due to the formation of a polymerization film on the surface of the fiber (Rippon 1984; Jocic et al. 2005; Kitkulnumchai et al. 2008).

β-Cyclodextrin

Cyclodextrins are torus-shaped cyclic oligosaccharides, built from six to eight ($\alpha = 6$, $\beta = 7$, $\gamma = 8$) D-glucose units. They form a dye-cyclodextrin complex and the complex can improve the dyeability of textile fibers. Among three cyclodextrins, β-cyclodextrin possesses better complexation efficiency than others. β-cyclodextrin forms an inclusion complex with dye molecules due to a slight bathochromic shift of the absorption maxima of the dye molecule. β-cyclodextrin is also used as retarding agents during dyeing of the wool fibers. Sulphonated β-cyclodextrin is fixed with wool fiber by the pad → dry → cure method and forms electrovalent bonds under the acidic condition and sulphonation modified β-cyclodextrin act as deep dyeing promoters for wool fabrics (Ammayappan and Jeyakodi Moses 2009b).

ColorZen

Cotton fibers are pretreated with a nontoxic chemical called ColorZen and the pretreated cotton fibers are spun into yarn and woven into fabric and then dyed with a conventional reactive dyeing process. ColorZen-treated cotton consumes 90 % less water than a conventional dyeing procedure so that there is an energy saving of 75 % and one third of the time. OekoTex Standard 100 approved ColorZen LLC and stated that ColorZen is 100 % free of harmful substances and safe for use on products for babies through adults (Anon 2015a).

Enzymes

Enzymes are used in the textile industry for the development of environmentally friendly preparatory processes as well as dyeing processes at lower temperature. Pectinases, cellulases, proteases, peroxidases, and lactases can replace insistent chemicals used in dyeing. Protease enzyme treatment improves the diffusion of dye molecules inside the wool fiber and increases the adsorption rate by decreasing the apparent activation energy for the dye. Commercial alkaline and neutral protease enzymes are mainly used for improvement in dyeability of selective woolen textiles. The lanazym process is based on purely enzymatic treatment for wool fabric with Perizym-AFW and it improves dyeability and washing fastness. Wool tops/knitted fabric treated with ammonia in the presence of sodium chloride, followed by an enzymatic treatment not only enhance the dyeing behaviour, but also shrink resistance. Similarly cellulase enzyme is used to enhance the brilliancy of the shade for cellulosic textiles (Lee et al. 1983; Cavaco-Paulo and Gubitz 2003).

Quaternary Ammonium Compounds

During dyeing of cotton fiber with direct or reactive dyes, electrolyte (NaCl or Na_2SO_4) are added to reduce the negative zeta potential present on the surface of the fibers. The addition of salts can increase in the BOD/COD of the effluent. Reactive quaternary ammonium compounds have been introduced to reduce the addition of electrolytes for dyeing of reactive dyes with cellulosic fiber. Glycidyl trimethylammonium chloride, N,N-dimethylaze-tidinium chloride, and N-methylol acrylamide are the compounds used to provide cationic sites in the cotton fiber (Burkinshaw 1989; Burkinshaw 1990; Lei and Lewis 1990; Lewis and Lei 1989).

7.1.3 Substrate

Textile products can be dyed in various forms including loose stock, sliver, top, yarn, fabric, garment, and nonwoven. Different factors such as production line, end product, potential of the dye house, and manpower decide the options for the dyeing operation in relation to the textile processes (Madaras et al. 1993; Ingamells 1993).

Loose Fiber

Loose fiber dyeing is mainly employed for the production of mélange yarn: loose stock is mainly dyed on the principle of circulating the liquor through the stationary material. Dyeing at the loose stock, sliver, and top stages can be applicable to dyes with poor leveling properties. However, this dyeing methodology takes too long with low production.

Yarn

Fine single yarn, knitting yarns, and high-twist fancy yarns are dyed in the form of cross-wound packages. Because there is no subsequent dyeing processing steps involved, the dyer must ensure level dyeing and it is the main concern in this dyeing method.

Fabric

Woven or knitted fabric is preferred for continuous dyeing and bulk production. Dyers can select dye in a wide range both in terms of leveling property and dyeing machine. Either the fabric is transported through the liquor (winch/jigger) or both the liquor and fabric are moved in opposite directions in the machine (jet dyeing machine). Open width fabric can be dyed in the jigger or pad-batch method, whereas the rope form can be dyed in a jet/winch dyeing machine. R&D work has been carried out to perfect a soft flow dyeing machine to reduce the energy, water, dye, and chemicals.

Garment

Garment dyeing is generally preferred for fancy effect and it requires more amounts of dyes, dyeing time, and skilled manpower. Automatic garment dyeing is carried out in a paddle/drum machine and used for dyeing house, fully fashioned garments, pullovers, and hats. Both material and dye solution are moved by means of the paddle blade/perforated drum.

Nonwoven

It is difficult to dye nonwoven products, inasmuch as these products can deform their structures. Nonwoven is usually dyed manually in a pot using a variety of colors.

7.2 Dyes

Selective dyes can have demerits such as hydrolysis at high temperature, low exhaustion rate, and poor fastness properties and these properties also reduce the sustainability. Research work has been carried out to enhance the exhaustion, fixation, and fastness properties.

7.2.1 Modified Dyes

For dyeing of 1 kg of cotton fabric with reactive dyes, up to 100 L of water are required in the conventional dyeing system. In addition, unfixed/hydrolyzed reactive dyes generally cause a reduction in fastness properties and require more rinsing treatment. M/s. Huntsman Textile Effects introduced a new tri-reactive dye called AVITERA–SE in the presence of ERIOPON LT (Clearing additive for reactive dye) and it requires 15–20 L of water to dye 1 kg of cotton fabric. They also introduced a clearing additive for hydrolyzed and unfixed reactive dye called

Table 4 Energy required for dyeing process and washing-off for 1 kg of cotton fabric

Dyeing System	Water (L)	Energy (kg steam)	CO_2 emission (kg electricity)	Processing time (h)
Hot dyeing system	60–100	9.0	35.0	9.0
Conventional warm dyeing system	40–80	6.5	2.5	7.0
Best available technology	30–40	3.9	1.5	5.5
Huntsman AVITERA SE and ERIOPON LT	15–20	1.7	0.65	4.0

ERIOPON® LT, which requires less than 60 °C rinsing bath to obtain the required colorfastness. The saving in water and energy by the Huntsman process in comparison with the best available technology (BAT) for reactive dyes is given in Table 4 (Lewis 2014; Anon 2015b).

7.2.2 Natural Dyes

Natural dyes have been used by humans to impart color to textile clothing since ancient times. Some people still use natural dyes for coloration of fashion textiles due to their distinct color, uniqueness, functionality, and ease of application. Coloration of textiles with natural dyes in the presence of biomordants could be one of the alternative sustainable and quality dyeing methods (Ammayappan and Seiko Jose 2015).

7.2.3 Dyes with Excellent Fastness Properties

Modified vat dyes for cellulosic textiles, 1:2 metal complex dyes for woolen textiles, leveling acid dyes for silk textiles, and bifunctional reactive dyes for cellulosic textiles give very good to excellent fastness properties. The selection would be based on the cost of dyeing and the end use (Lewis 2014).

7.3 Dye Bath Additives

Ionic dye molecules exist as aggregated molecules and their diffusion inside the fiber polymer depends on the size of aggregation. If the radii of dye aggregates exceed the pore size of the fiber, dyeing time will be long. Dye bath additives are either ionic or nonionic in nature and can be added in the dyeing bath to keep individual dye molecules through stabilization. Dye bath additives such as the exhausting agent, fixing agent, protecting agent, and leveling agent are used to improve the dyeing rate as well as uniformity of dyeing. They are anionic, cationic, amphoteric, and nonionic in nature (Ammayappan et al. 2011).

7.3.1 Protecting Agents

Dyeing at high temperature of woolen textiles can lead to adverse effect on downstream processing as well as product performance. JA. Rippon of CSIRO, Australia developed an amphoteric surfactant called Sirolan-LTD, used to remove the lipids on the surface of the wool fiber. By using this protective agent, dyeing of woolen textiles can be carried out at 80–90 °C for all classes of dyes with excellent fastness properties. After dyeing, the residual substance biodegrades easily and leaves no toxic residues (Lewis and Rippon 2013)

7.3.2 Leveling Agents

Dye migration and fixation inside the fiber matrix decide the solidity of the shade. If the affinity of the dye molecules towards the fiber is high, there is a possibility of rapid diffusion and it may lead to uneven dyeing. Leveling agents are used to improve the leveling of the dye molecules by controlling the dyeing rate and improving the uniform distribution of dye molecules. It is mainly used for dyeing of woolen textiles with metal complex and high molecular weight acid dyes (Ammayappan 2009).

7.3.3 Liposomes

Liposomes are spherical synthetic layers of phospholipids, which are formed like closed vesicles with an aqueous core, ranged from 10 to 100 nm in diameter. They consist of hydrophilic phosphate and choline groups, and a hydrocarbon hydrophobic part. The hydrophobic part is used as a carrier for the dye and can be easily absorbed by the wool fiber. The interaction between the lipid of liposomes and cell membrane complex of wool leads to uniform diffusion of dye molecules inside the fiber at low temperature (85 °C). Commercially liposomes are used as a textile auxiliary in wool dyeing which has been adapted by some textile industries. This process also has additional benefits including improved smoothness with retention of mechanical properties with a clear reduction in the dye house effluents (Ammayappan 2009; De La Maza et al. 1997).

7.3.4 Reverse Micelle Medium

The reverse micelle system is used to dye protein fiber with anionic dyes in a nonaqueous dyeing medium. Acid dye is dissolved in the water-pool in the presence of the organic compound called decamethyl cyclopentasiloxane (D5) to form a dye solution/nonaqueous medium emulsion. This system has an organic solvent as a continuous phase and dyes with hydrophilic fiber in a nonaqueous medium (Sawada and Ueda 2003; Song 2007). This process can reduce the consumption of water for dyeing.

7.4 Dyeing Methods

7.4.1 Air Dyeing

Air dyeing was developed by M/s. Colorep, a California-based company and this process used up to 95 % less water and up to 86 % less energy in comparison with conventional dyeing methods. This process resembles transfer printing of polyester fiber with disperse dyes in which air is used instead of water for the penetration of the disperse dyes into the fiber. The disperse dye is applied on some special types of paper and the dye is transferred to the fabric surface by the application of heat. They reported that this technology could save around 75 gallons of water for dyeing of a single pound of fabric (Otsuki and Raybin 2015; Anon 2015c).

7.4.2 Digital Printing

Similar to the AirDye principle, recently digital printing has been used for fashion garments in which designs can be printed on the fabrics, so that there is a reduction in water usage up to 95 % and energy up to 75 % in comparison with a conventional printing process (Anon 2015d).

7.4.3 Hand Dyeing of Natural Dyes

Eindhoven-based designers Renee Mennen and Stefanie van Keijsteren introduced a traditional hand-dyeing method to develop fashion clothing in which they have used selective natural dyes with eco-friendly mordants for the development of a monochrome rainbow of red colors on woolen and cotton textiles (Anon 2015d).

7.4.4 Dyeing at Inert Atmosphere

M/s Master S R L, Italy, has introduced an environmentally friendly dyeing process called Genius2 for indigo and sulfur dyes on cellulosic textiles in a nitrogen atmosphere. This process kept the leuco vat/sulphur dye exists in nano stage and a special diffusion/fixation unit is used to enhance the fixation of dye molecules. Similarly they also developed a continuous dyeing machine for denim fabrics called IndigoGenius, in which indigo/sulphur dyes can be dyed at nitrogen atmospheric conditions. This process shows high color yield and reduction (up to about −80 %) in consumption of sodium hydroxide and hydrosulfite in comparison with conventional dyeing (Anon 2015e).

7.4.5 Nano-Dye Process

M/s. Nano-Dye Technologies, Inc., United States, introduced a process called Nano-Dye. It is a continuous dyeing system in which dye molecules are kept in individual nano stages as well as modifies the charge of the cotton fiber to cationic charge. These modifications lead to the highest dye exhaustion of reactive dyes with a low amount of additives, so that it consumes 90 % less energy, 70 % less wáter, and almost nearly zero solid waste than a conventional dyeing process (Anon 2015f). This process is in the developmental stage and will be implemented in late 2015.

7.4.6 Super-Critical Carbon Dioxide Dyeing

DyeCoo, a Dutch-based Company, developed industrial super-critical carbon dioxide (scCO$_2$) dyeing technology by using carbon dioxide to act as a fluid similar to water through temperature- controlled pressure chambers. In this method, the dyestuff is first dissolved in scCO$_2$ and is then transferred to the substrate. The dissolved dye gets subsequently absorbed and diffused inside the fiber due to the swelling effect of scCO$_2$. After dyeing, residual dyes and scCO$_2$ can be recovered. It is observed that the equilibrium sorption of the dye on fiber is increased with increase in pressure and temperature (Schmidt et al. 2003)

7.4.7 Microwave-Assisted Dyeing

Microwave is a form of electromagnetic radiation with wavelengths ranging from 10^{-3} to 1 mm, with frequencies between 300 MHz and 300 GHz (Yurika 2005). Generally it is used to heat the bulky woolen cheese rapidly in a uniform manner and consequently in dyeing inasmuch as it saves dyeing time, wáter, and chemicals. Haggag et al. (2014) inferred that microwave-assisted dyeing of cellulosic textile with reactive dyes reduced the dyeing time, salt, and alkali consumption around 90 %, 75 %, and 20 %, respectively, without deteriorating the fastness properties. Zhao and He (2011) treated woolen fabric in a microwave oven at 2450 MHz, 250–1000 W for 30–180 s and inferred that microwave-modified wool fiber showed an improvement in diffusion behaviour of dye molecules without deteriorating its mechanical properties.

7.4.8 Ultrasound-Assisted Dyeing

Ultrasound consists of high-frequency (>18 kHz) inaudible oscillations. A current of 50–60 Hz alternate current is converted into a high-frequency electrical energy by the generator. High-frequency electricity is fed to the transducers and is transformed into mechanical vibrations. The transducer system vibrates longitudinally, transmitting waves into the liquid medium. In liquid as the ultrasound waves propagate, they form microscopically small bubbles. These bubbles expand and finally during the compression phase they collapse violently and generate shock waves. The waves are responsible for ultrasonic effects observed in a liquid/liquid system (Ammayappan 2009). Ultrasound influences the dyeing process by dispersing dye molecules into individual form, increasing activation energy of dye molecules, and making rapid diffusion inside the fiber structure so that it enhances the rate of dyeing. Yu et al. (2010) studied ultrasound-assisted dyeing of wool fiber with reactive and acid dyes at 50–55 °C and found that this process is more economical than a conventional dyeing process in terms of energy, color yield, and product quality. Ferrero and Periolatto (2012) compared the dyeing of wool fiber with and without ultrasound. It is reported that ultrasound-based dyeing exhausted acid dyes at 60–80 °C, whereas a conventional method required 90–95 °C for equivalent exhaustion.

7.4.9 Electrochemical Dyeing

Electrochemical dyeing is introduced to replace conventional dyeing of cellulosic textiles with vat or sulphur dyes in which electrochemical energy is used to reduce the dyes instead of conventional chemical reduction with sodium sulphide or sodium hydrosulphite. Anbu Kulandainathan et al. (2008) carried out electrochemical dyeing of vat dyes with a very low concentration of sodium dithionite by using iron-deposited graphite as the cathode and reported that the color intensity and washing fastness of the dyed fabrics were found to be equal with conventionally dyed fabrics. Bechtold et al. (2008) reduced a sulphur dye in a multicathode electrolyzer and observed that there is a positive correlation between redox potential and color depth of the dyed samples. Babu et al. (2009) performed electrochemical dyeing of vat dyes with ferrous/ferric complexes coupled with Ca complexes instead of hydrosulphite. Ferric triethanol amine (TEA) complex in alkaline medium is a well-known redox mediator and can be used to reduce the vat dyes in this electrochemical process. After dyeing, the solution can be reused by filtering the dyes in the complexes.

7.5 Machinery Development

Jet dyeing machines are used to dye hosiery fabrics and work on the principle of accelerating water through a nozzle to transport fabrics through the machine. Recently low liquor ratio jet dyeing machines have been introduced, in which low friction Teflon internal coatings and advanced spray systems are used to speed up the dyeing and rinsing process. Similarly ultralow liquor ratio jet dyeing machines operate at a 6:1 liquor ratio in comparison with the LLR jet dyeing machine (8:1). M/s. Cleantech Solutions International, Inc., China has developed a dyeing machine which uses both air flow and fluid flow in the dyeing process with a series of nozzles, cloth wheels, and cloth spreaders. This dyeing machine can use 60–70 % less water, 50 % less additives, 30 % less power, and 40–50 % less steam than a conventional dyeing machine. Due to less consumption of chemicals, it also shortens the dyeing time by 1–2 h by ensuring good color evenness (Anon 2015g).

M/s. Then Maschinen GmbH, Germany, introduced the Airflow® Lotus 200 dyeing machine in which they replaced the dye liquor with moisture-saturated air flow for transporting piece goods in jet-dyeing machines with low MLR. It leads to high transportation, energy savings of up to 40 %, and 25 % reduction in processing time in lowest water levels (Anon 2015h). A new airflow technology, aerodynamic, has been developed by M/s. Fong's Europe GmbH, Germany and dyed cotton-based textiles in both woven and hosiery with less than 4:1 MLR. It is reported that total water consumption by a conventional jet dyeing process ranged 40–100 L whereas aerodynamic technology requires 38–45 L of water/kg for cotton textiles (Anon 2015i). Similarly, M/s. Thies GmbH, Germany, has developed dyeing machinery called Thies 2000 iMaster H_2O for cotton and cotton-blended hosiery fabrics with 1:3.7 MLR. It is reported that it consumed 0.28 kW power and 17–50 L of water/kg of fabric and can save up to 50 % energy as well as water consumption (Anon 2015j).

7.6 Management

Other than development in the dyeing machinery and method, respective management is also responsible for sustaining the dyeing process by executing the standard protocols to save energy and chemicals (Park and Shore 2009; Teli 1996). They should make the proper arrangements to explore the sustainability of the dyeing process among all staff members and also encourage the involvement of staff members to follow the rules for sustainability by giving incentives and awards.

7.6.1 Right-First-Time Dyeing

The main objective of the right-first-time dyeing (RFTD) is to increase the productivity of a processing unit by achieving the desired shade on a product at the first attempt in the right time by selecting the proper dye/dyeing method/machinery. By RFTD, every dyeing industry can save dyeing time, increase profit, and reduce the dyeing cost as well as reduce the pollution load (Park and Shore 2009).

7.6.2 Efficient Practices

Efficiency of the dyeing process can be improved by the automation of the unit processes in the dyeing unit. It also leads to reducing the pollution or waste and energy without deteriorating the strength, but improves the fastness properties of clothing in consistent ways (Perkins 1991; Thiry 2011). Some of the potential and efficient practices are given below:

1. Dye bath reuse: It is applicable for dark shades in batch process and unfixed dyes of the previous bath can be reused in order to reduce the effluent load.
2. Continuous dyeing for knits: Based on the capacity and the customer requirement, hosiery fabrics can be dyed continuously in order to reduce the dyeing cost.
3. Automated color mix kitchen: Dye can be dosed in the dyeing machine in a controlled manner so that waste in dye can be reduced.
4. Automated chemical dosing: Monitoring the dye bath additives can greatly improve the dyeing quality with good reproducibility of the process and maintaining the process fortitude.
5. Countercurrent washing: Water is the main source for a dyeing unit; if the concurrent washing plant can be implemented it can conserve 20–30 % of water.

7.6.3 Design-Stage Planning

The planning stage for new processes and products is essential because it offers the opportunity to design sustainable dyeing technologies. It can be achieved by deploying an experienced R&D group comprising scientists, technologists, dyers, and management staff. This group can survey the latest trends in dyeing and conduct brain-storming sessions with all categories of people to execute the sustainable dyeing process (Jeyakodi Moses and Ammayappan 2006).

Design-Stage Planning for Processes

Design-stage planning for processes focuses on arranging the dyeing procedure in a proper way to reduce the pollution load. It can be executed by examining the existing dyeing process at the fundamental design level and rectifying/modifying the demerits in order to improve the dyeing quality. For example, dyeing cost is a major role in planning of the dyeing industry and so management must consider the following dyeing sequence in order to select the proper dyeing process for their material.

Solubilised vat > Vats > Bi-functional reactive > Metallised direct > Azoic and unmetallised direct > Liquid sulphur (for cellulosic textiles)

1:2 Metal complex > Reactive > 1:1 Metal complex > Acid dye > Direct dye (for protein fiber textiles)

Design-Stage Planning for Products

Today's consumer expectations are environmentally friendly, cost-worth, and well-designed products. Each industry management must adopt an innovative supply chain for delivering the product as per the customer's requirements including environmental aspects. For example, shades and colors should be selected that use the most environmentally benign dyes.

7.6.4 Dissemination of Standardized Methodology

Every successful dyeing technology has ease of adaptability from one to another corner of the world in terms of repeatability, cost, economics, and eco-friendliness. The standardization protocol and reporting format of the dyeing process can be used to transfer its information. This type of dissemination surely produces successful results and it requires minimal cost and effort.

7.6.5 Biofiltering of Wastewater

Wastewater treatment is considered a supportive methodology for sustaining a dyeing process indirectly. A comprehensive technology called a "sequencing batch biofilter granular reactor" has been developed in which it oxidizes the textile dyes by ozone treatment followed by a biofiltration. The treated effluent holds more microorganisms than traditional technologies and produces 80 % less sludge than conventional filters (Lotito et al. 2011).

7.6.6 Smart Tailoring

Reduction of the wastage in garment manufacturing can indirectly sustain the dyeing process. Direct Panel on Loom (DPOL) or Smart Tailoring technology was created by Indian designer Siddhartha Upadhyaya, and can be used to manufacture a garment by well-planned weaving, fabric cutting, and patterning in a consequent manner so that this process minimizes garment waste as well as helps in saving energy and water by 70–80 % (Anon 2015d).

7.6.7 Eco-Labeling

Eco-certification of a garment can also suggest that management select suitable material, dyes, process, and machinery in order to provide credible assurance to the consumer. It also leads to gain universal recognition, enabling processors and manufacturers to supply a universally acceptable product. There are two types of eco-labeling systems: the first type is privately owned certification agencies such as Oeko-Tex 100 and second type is association based agencies such as the European Union Eco-label (Ammayappan and Jose 2015).

7.7 Education

A long-term approach to sustain the dyeing process can be achieved by formalized employee education. Education programs are more general and less job-oriented than training programs. Several specific topics have been documented in the literature, including:

1. In-depth understanding of chemistry, reaction kinetics of dye, and dyeing machinery among technical executives is essential to foster awareness of sustainability.
2. Establishment of corporate-level work groups to develop and distribute information about sustainable dyeing processes.
3. Internal training and education through in-house newsletters is an effective way to communicate information and educate employees.
4. Several useful external training and education mechanisms including conferences, equipment and trade shows, in-plant courses by outside experts or plant technical personnel, and promoting correspondence courses from textile colleges to update their knowledge on sustainable dyeing.

8 Future Perspectives

Adaption of bulk-scale production of a technology in any sector depends on its public demand, feasibility, economy, and government policy. The natural dye industry has existed as a small-scale sector for a long time. After introduction of synthetic dyes and their ease of application protocol, many governments promoted the dyeing industry based on synthetic dyes because it contributes a major share of the GDP as well as employment for a country. Initially, there was no restriction in the testing of the dyes and discharge of the dye effluent. Subsequently the textile world sent out an alarm to check the quality of the synthetic dyes in terms of eco-friendliness during the 1990s. After imposing the ban on selective synthetic dyes, the dyeing sector undertook many innovative methods to reduce the water, energy, and chemicals used in processing cotton textiles by 50 %; for example, during the 1990s 130–200 L of water/kg of fabric were used to dye 1 kg of cotton textile which was reduced to 65–70 L in the 2000 s, and chemical and energy consumption were reduced to 40 and 50 %, respectively.

In the twenty-first century, there are many restrictions starting from raw material procurement to certification of the finished goods. In practice those rules can be followed in the developed countries, but the adaption of those rules and regulation may be varied in the developing countries such as India, Bangladesh, and Cambodia, where the majority of the textile dyeing industries are running. Most of the dyeing industries did not follow those rules due to the involvement of huge investment in costly machinery, advanced purification systems, and modern effluent treatment plants that led to the closure of many dyeing industries. However, to meet the global demand for clothing, they move their dyeing operations to another city without following the rules for sustainability. It is the responsibility of all members of the dyeing sector to achieve a sustainable impact by having clear management of the supply chain of fashion textiles. This is the correct time to rationalize the tactical plans by collaborating in all possible ways to make the dyeing industry sustainable for our future generations.

9 Conclusion

Dyeing was introduced in our civilization as an art for imparting colors on textile clothing in order to enhance the dignity of the people. Natural dye based coloration was an art and executed by skilled artisans for many centuries. After the invention of synthetic dyes and industrialization in the nineteenth century, consumption of synthetic dyes for coloration progressed rapidly. Selection of a dye for coloration is based on the cost of the dye and consumer demands. Coloration of textile materials with synthetic dyes has increased due to the increase in population up to the 1990s. After exploration of carcinogenicity of synthetic dyes and introduction of the eco-label, consumers became aware of the eco-friendliness of their products. The eco-label creates not only awareness of green textile; it also acts as a starting point for the sustainable dyeing processes and reduces the pollution load.

Researchers and academicians are constantly working on sustaining technology that can reduce the pollution load in the textile industry, because conventional dyeing consumes more energy and water. Natural textile fibers are heterogeneous and have natural impurities, thus it is always a great challenge for scientists to develop a suitable technology for an energy-efficient, cost-effective, and particularly sustainable dyeing methodology. Adaption of a sustainable technology from laboratory to industry is a deciding role and it depends on so many factors such as their economy, ease of operation, and technological know-how about the dyeing. Development of quality and green textile products by adapting those sustainable technologies irrespective of their cost can surely lead to survival of natural resources, which is the basic need of future generations.

References

Ammayappan L (2009) Innovative technologies for dyeing of woolen products. In: Karim SA Shakyawar, DB Anil Joshi (eds) Wool technology: innovations in wool production and technologies for value addition. Agrotech Publishing Academy, Udaipur, India. pp. 323–335

Ammayappan L (2013a) Application of enzyme on woolen products for its value addition: an overview. J Text Apparel Manage 8(3):1–12

Ammayappan L (2013b) Eco-friendly surface modifications of wool fiber for its improved functionality: an overview. Asian J Text 3(1):15–28

Ammayappan L, Jeyakodi Moses J (2009) An overview on application of cyclodextrins in textile product enhancement. J Text Assoc 70(1):9–18

Ammayappan L, Muthukrishnan G, Saravana Prabhakar C (2003) A single stage preparatory process for woven cotton fabric and its optimization. Manmade Text India 46(1):29–35

Ammayappan L, Nayak LK, Ray DP et al (2012) Plasma treatment on textiles. Asian Dyer 8(6): 34, 37–40

Ammayappan L, Seiko Jose (2015) Functional aspects, eco testing and environmental impact of natural dyes. In: Muthu SS (ed) Handbook of sustainable apparel production. CRC Press, Boca Raton, pp. 333–350

Ammayappan L, Shakyawar DB, Gupta NP (2011) Optimization of dyeing condition for wool/cotton union fabric with direct dye using Box-Behnken Design. Fibers Polym 12(7):957–962

Anbu KM, Kiruthika K, Christopher G et al (2008) Preparation of iron-deposited graphite surface for application as cathode material during electrochemical vat-dyeing process. Mater Chem Phys 112:478–484

Angelini LG, Pistelli L, Belloni P et al (1997) Rubia tinctorum a source of natural dyes. Ind Crops Prod 6:303–311

Anon (2015a) What is ColorZen? http://www.colorzen.com/. Accessed 25 July 2015

Anon (2015b) http://www.textileworld.com/Issues/2011/March-April/Dyeing_Printing_and_Finishing/Sustainable Dyeing And_Finishing Accessed 21 July 2015

Anon (2015c) http://www.airdyesolutions.com/. Accessed 29 Aug 2014

Anon (2015d) Website http://www.treehugger.com/sustainable-fashion/10-awesome-innovations-changing-future-fashion.html. Accessed 12 Aug 2015

Anon (2015e) Technology. http://www.mastermacherio.it/. Accesses 12 Aug 2015

Anon (2015f) Keep up with NANO-DYE TECHNOLOGIES, INC. https://www.linkedin.com/company/nano-dye-inc. Accessed 24 July 2015

Anon (2015g) Our Products. http://www.cleantechsolutionsinternational.com/product.php. Accessed 15 Aug 2015

Anon (2015h) recent developments in dyeing. http://www.textileworld.com/Issues/2010/March-April/Features/Recent_Developments_In_Dyeing.html. Accessed 12 Aug 2015

Anon (2015i). THEN-AIRFLOW® SYNERGY-The consequence of piece dyeing. http://www.fongs.eu/assets/templates/fongs/img/pdf/THEN/THEN_SYNERGY/THEN_SYNERGY_Eng.pdf. Accessed 16 Aug 2015

Anon (2015j) Thies GmbH and Fong's Europe GmbH. http://www.textileworld.com/Articles/2011/March/March_April_issue/files/DPFTable1.pdf. Accessed 10 Aug 2015

Atav R (2014) The use of new technologies in dyeing of proteinous fibers. http://dx.doi.org/10.5772/53912. Accessed 15 Aug 2015

Atav R, Yurdakul A (2011) Low temperature dyeing of plasma treated luxury fibers. Part I: results for Mohair (Angora goat). Fibers Text East Eur 19(2):84–89

Babu F, Senthil Kumar K, Anbu Kulandainathan RM et al (2009) Ferric-oxalate-gluconate based redox mediated electrochemical system for vat dyeing. J App Electrochem 39(7):1025–1031

Bae S-H, Motomura H, Morita Z (1997) Diffusion/adsorption behaviour of reactive dyes in cellulose. Dyes Pigm 34:321–340

Bechtold T, Turcanu A, Schrott W (2008) Electrochemical reduction of CI sulphur black 1-correlation between electrochemical parameters and colour depth in exhaust dyeing. J Appl Electrochem 38:25–30

Beevers RB, McLaren KG (1974) The effect of low doses of Co60 gamma radiation on some physical properties and the structure of wool fibers. Text Res J 44(12):986–994

Bertea A, Bertea AP (2008) Decolorisation and recycling of textile wastewater (in Romanian), Performantica edn. Iasi, Romania

Bide M (2014) Sustainable dyeing with synthetic dyes. In: Muthu SS (eds) Roadmap to sustainable textiles and clothing. Springer, Singapore, pp 81–107

Bird CL, Boston WS (1975) The theory of coloration of textiles. Dyers Company Publications Trust, Yorkshire

Broadbent AD (2001) Basic principles of textile coloration. Society of Dyers and Colourists, England

Burkinshaw SM, Lei XP, Lewis DM (1989) Modification of cotton to improve its dyeability, Part I: Pretreating cotton with reactive polyamide-epichlorohydrin resin. J Soc Dyers Colour 105:391–398

Burkinshaw SM, Lei XP, Lewis DM et al (1990) Modification of cotton to improve its dyeability, Part 2: Pretreating cotton with Thiourea derivate of polyamide-epichlorohydrin resins. J Soc Dyers Colour 106:307–315

Cavaco-Paulo A, Gubitz G (eds) (2003) Textile processing with enzymes. Woodhead Publishing, England

Cegarra J (1998) Determination of the migratory properties of direct dyes. J Soc Dyers Colour 73(8):375–381

Cho KR (1999) Studies on natural Dyes 11-Dyeing properties of cochineal colors for wool fabrics. J Korean Soc Dyers Finish 11:39-49

Correia VM, Stephenson T, Judd SJ (1994) Characterization of textile wastewaters: a review. Environ Technol 15:917–992

De La Maza A, Coderch L, Parra JL et al (1997) Multi lamellar liposomes including cholesterol as carrier of a 1:2 metal complex dye in wool dyeing. Text Res J 67(5):325–333

Dos Santos AB, Cervantes FJ, Van Lier JB (2007) Review paper on current technologies for decolourisation of textile wastewaters: perspectives for anaerobic biotechnology. Bioresour Technol 98(12):2369–2385

EPA (1997) Profile of the textile industry. Environmental Protection Agency, Washington, DC

EWA (2005) Efficient use of water in the textile finishing industry. European Water Association, Brussels

Ferrero F, Periolatto M (2012) Ultrasound for low temperature dyeing of wool with acid dye. Ultrason Sonochem 19:601–606

Forgacs E, Cserháti T, Oros G (2004) Removal of synthetic dyes from wastewaters: a review. Environ Int 30(7):953–971

Haggag K, El-Molla MM, Mahmoued ZM (2014) Dyeing of cotton fabrics using reactive dyes by microwave irradiation technique. Indian J Fibre Text Res 39(4):406–410

Ingamells W (1993) Colour for textiles: a user's handbook. Society of Dyers and Colourists, England

Jeyakodi Moses J, Ammayappan L (2006) Growth of textile industry and their issues on environment with reference to wool sector. Asian Dyer 3(3):61–68

Jocic D, Vílchez S, Jocic D, Topalovic T et al (2005) Chitosan/acid dye interactions in wool dyeing system. Carbohydr Polym 60:51–59

Johnson A (1989) The theory of coloration of textiles, 2nd edn. Society of Dyers and Colorists, England

Kitkulnumchai Y, Ajavakom A, Sukwattanasinitt M (2008) Treatment of oxidized cellulose fabric with chitosan and its surface activity towards anionic reactive dyes. Cellulose 15:599–608

Lee SB, Kim IH, Ryu DDY et al (1983) Structural properties of cellulose and cellulase reaction mechanism. Biotechnol Bioeng 25(1):33–51

Lewis DM, Lei XP (1989) Improved cellulose dyeability by chemical modification of the fiber. Text Chem Color21:23–29

Lei XP, Lewis DM (1990) Modification of cotton to improve its dyeability,Part 3: Polyamide-epichlorohydrin resinsand their ethylendiamine reaction products. J Soc Dyers Colour 106:352–356

Lewis DM (2014) Developments in the chemistry of reactive dyes and their application processes. Coloration Technol 130:382–412

Lewis DM (1992) Wool dyeing. Society of Dyers and Colourists, England

Lewis DM, Rippon JA (eds) (2013) The coloration of wool and other keratin fibres. Wiley, England

Lewis DM, Lei XP (1989) Improved cellulose dyeability by chemical modification of the fiber. Text Chem Color 21:23–29

Lotito AM, Iaconi CD, Fratino U et al (2011) Sequencing batch biofilter granular reactor for textile wastewater treatment. New Biotechnol 29(1):9–16

Madaras GW, Parish GJ, Shore J (eds) (1993) Batch wise dyeing of woven cellulose fabrics: a practical guide. Society of Dyers and Colourists, England

Mahapathra NN (2011) Dyeing technologies for future. Colourage 108(6):49–52

Marsh JT (1941) Mercerizing, 1st edn. Chapman and Hall, London

Mendez J, Gonzlez M, Lobo MG et al (2004) Color quality of pigments in cochineals Dactylopius coccus Costa: Geographical characterization using multivariate statistical analysis. J Agric Food Chem 52:1331–1337

Moustafa S (2008) Environmental impacts of textile industries. Process analysis of textile manufacturing. UNESCO-IHE, Delft

Mussig J, Stevens C (eds) (2010) Industrial applications of natural fibres: structure, properties and technical applications. Wiley, New York

Noah AO, Martins CMOA, Braimah JA (1986) The effect of electrolytes on direct dyes for cotton. J Appl Polym Sci 32(7):5841–5847

O'Neill C, Hawkes FR, Hawkes DL et al (1999) Colour in textile effluents—sources, measurement, discharge consents and simulation: a review. J Chem Technol Biotechnol 74(11):1009–1018

Otsuki J, Raybin P (2015) AIRDYE® Environmental Profile-Life Cycle Assessment. http://www.airdyesolutions.com/uploads/AirDye_EPDv2b_091109.pdf. Accessed 29 Aug 2015

Park J, Shore J (2009) Evolution of right-first-time dyeing production. Color Technol 125(3):133–140

Perkins WS (1991) A review of textile dyeing processes. Am Assoc Text Chem Color 23(8):23–27

Peters RH, Vickerstaff T (1948) The adsorption of direct dyes on cellulose. Proc R Soc Lond Ser A Math Phys Sci 192:292–308

Rippon JA (1984) Improving the dye coverage of immature cotton fibers by treatment with Chitosan. J Soc Dyers Colour 100(10):298–303

Robinson T, McMullan G, Marchant R et al (2001) Remediation of dyes in textile effluent: a critical review on current treatment technologies with a proposed alternative. Bioresour Technol 77(12):247–255

Rys P, Zollinger H (1972) Fundamentals of the chemistry and application of dyes. Wiley-Interscience, London

Sawada K, Ueda M (2003) Dyeing of protein fiber in a reverse micellar system. Dyes Pigm 58:99–103

Schindler WD, Hauser PJ (eds) (2004) Chemical finishing of textiles. Woodhead Publishing, England

Schmidt A, Bach E, Schollmeyer E (2003) The dyeing of natural fibers with reactive disperse dyes in supercritical carbon dioxide. Dyes Pigm 56:27–35

Shamey R, Zhao X (2014) Modelling, simulation and control of the dyeing process. Woodhead Publishing, England

Shore J (1995) Cellulosics dyeing. Society of Dyers and Colourists, England

Song XY (2007) New dyeing technologies using reactive dyes [part I]. China Text Leader 9:108–110

Teli MD (1996) New developments in dyeing process control. Indian J Fibre Text Res 21(1):41–49

Thiry MC (2011) Staying alive: making textiles sustainable. http://www.aatcc.org/wp-content/uploads/2015/03/Sustain1111.pdf. Accessed 16 July 2015

Trotman ER (1984) Dyeing and chemical technology of textile fibers, 6th edn. Edward Arnold, London

Venkataraman K (1971) The chemistry of synthetic dyes Vol-VIII. Academic Press, London

Vijaraghavan NS (1999) Environmental unit in textile industry. Bhopal Sci Tech Entrep 7:3–9

Wei Q (2009) Surface modification of textiles. Woodhead Publishing, England

Yu Y, Chen X, Zhu W (2010) Application of ultrasonic technique in low temperature dyeing of wool. J Text Res 13:70–72

Yurika Y (2005) Effect of microwave heating on dyeing. Seni Gakkaishi 63(6):40–48

Zhao X, He JX (2011) Improvement in dye ability of wool fabric by microwave treatment. Indian J Fibre Text Res 36(1):58–62

Developments in Sustainable Chemical Processing of Textiles

A. Arputharaj, A.S.M. Raja and Sujata Saxena

Abstract Chemical processing adds value to the textiles by improving aesthetics and imparting functional properties. It is usually carried out in the aqueous medium and thus requires a large amount of water. A number of chemicals and auxiliaries are employed in the process many of which are not biodegradable. Unused chemicals are discharged along with the process water as effluent which has to be treated at huge costs to make it comply with environmental regulations. Textile processing is energy intensive also as many treatments are carried out at elevated temperatures. Requirement of these inputs depends upon the nature of the fibre and machine used. As discharge and treatment of the aqueous effluent and unavailability of soft water required by the textile industry is the biggest challenge towards ensuring sustainability of the textile-processing industry, most of the developments in this field have tried to address these issues in various ways. This chapter analyses the key issues in the textile wet processing with special emphasis on the usage of dyes, chemicals, water, energy, carbon footprints, and problems associated with disposal of harmful chemicals to the environment. Research and development in sustainable processing using enzymes and natural products with better biodegradability have been discussed. Waterless technologies for textile processing with special citations of supercritical and plasma technology have been reviewed. Developments in dyes and dyeing for higher sustainability were critically analysed and alternatives for the source reduction at various processing stages have been explored. Social responsibility of different stakeholders for sustainable textile wet processing has also been discussed.

Keywords Sustainable processing · Textile processing · Carbon footprints · Eco-friendly processing · Waterless technologies · Source reduction · Social responsibility

A. Arputharaj (✉) · A.S.M. Raja · S. Saxena
Chemical and Biochemical Processing Division, ICAR-Central Institute
for Research on Cotton Technology, Adenwala Road, Matunga, Mumbai, India
e-mail: arajseeli@gmail.com

© Springer Science+Business Media Singapore 2016
S.S. Muthu and M.A. Gardetti (eds.), *Green Fashion*,
Environmental Footprints and Eco-design of Products and Processes,
DOI 10.1007/978-981-10-0111-6_9

218   A. Arputharaj et al.

1 Introduction

Chemical processing is the most important operation in the manufacturing of textile products. It not only adds value to the textile products, but also improves the comfort and aesthetic properties. Textile materials have undergone chemical processing since time immemorial. In all the years, the basic objectives of chemical processing, that is, colouring and finishing have not changed, but in recent times this field has expanded and diversified. The array of numerous fibres with different chemical nature such as cotton, wool, silk, polyester, nylon, and acrylic among others has greatly increased the complexity of the chemical processing. Chemical processing methods of textiles can be broadly classified into three major groups—cellulosic, protein, and synthetic origins (Ramesh Babu et al. 2007)—and the major steps involved for these groups are listed in Table 1.

Application of a particular unit operation in the actual processing is dependent upon the nature of the material (i.e., fiber, yarn, fabric) and the end use of the material (i.e., apparel or technical textiles). Figure 1 explains the input–output structure of the chemical processing industry. This process involves heterogeneous interaction between fibre polymers which are solids and chemicals that are mostly liquids or in aqueous solution form. Chemical processing of textiles requires a huge amount of water and energy. Water, which is left after the different processes is contaminated with residual dyes and chemicals that are harmful to the environment if not properly treated. The textile processing industry is considered one of

Table 1 Different stages in the processing of fibrous materials

Cellulosic	Protein	Synthetic
Desizing	Carbonization/degumming	Prewashing
Scouring	Scouring	Bleaching
Mercerization	Bleaching	Heat setting
Bleaching	Colouration	Colouration
Colouration	Finishing	Finishing
Finishing		

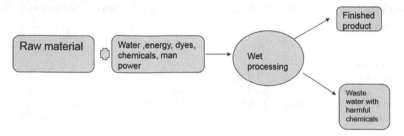

Fig. 1 Input–output structure of chemical processing industry

the major environment polluting sectors. It is also one of the biggest greenhouse gas (GHG) emitters on Earth. Environment-related issues in the chemical processing sector have been strongly felt the world over since the last decade of the previous century.

The German ban on azo dyes in the 1990s can be considered as the eyeopener for the textile processing industries about the environmental impact of different chemicals used in wet processing operations. Textile industries are triggered to implement green processes due to the introduction of new eco-labels and stringent environmental norms for the effluents by the governments. The growing number of green buyers worldwide is also driving the industries to adopt eco-friendly sustainable methods for processing. Textile processes require an enormous quantity of energy in the form of heat which leads to significant consumption of fossil fuels and increase in carbon footprints. Hence textile processing industries are the prime targets of the environmentalists in their crusade against pollution. This industry needs to satisfy the demands of stringent legislation and norms imposed by governments and other international bodies to comply with sustainable issues. As per the report of World Commission on Environment and Development (1987) sustainable development is defined as 'Development that meets the needs of the present without compromising the ability of future generations to meet their own needs.' This chapter analyses the important sustainability issues in textile wet processing such as usage of water, energy, and problems associated with usage and disposal of harmful chemicals to the environment. Research and development in sustainable processing and finishing using natural materials and enzymes which have better biodegradability have also been discussed. Waterless technologies such as supercritical carbon dioxide and plasma technology have been reviewed. Research and commercial developments taken place in dyes, mainly reactive dyes and dyeing including machinery developments, have been critically analysed. Alternatives for the source reduction using eco-friendly alternatives for various processing steps have been explored. Advantages and limitations of eco-labels and social responsibility of different stakeholders in the implementation of sustainable textile processing have also been discussed.

2 Current Practices and Sustainability Issues in Textile Processing

Sustainability issues in textile processing can be categorized into three major areas as follows.

- Usage of water
- Energy consumption and carbon footprints
- Pollution load and waste generation

2.1 Usage of Water

Water is the one of the main inputs in chemical processing industries and the textile wet processing sector ranks among the top 10 water-consuming industries. Water is used for various operations in wet processing as illustrated in Fig. 2. The main difference between textile processing and other industries is that, whatever water is used in the process, the final product is not carrying the water. Most of the water used in the processing is discharged as effluent which is contaminated with toxic dyes and chemicals. Reusing this water requires tedious and complex effluent treatments. It is estimated that approximately 50–200 L of water are required for the conversion of one kg of raw textile into finished product.

Cotton is the king of natural fibres and finds predominant use in the manufacture of apparel-based textiles. Cotton accounts for nearly 40 % share of the total global fibre consumption. Raw cotton fibres have to undergo a series of chemical processes to convert them into finished products. Table 2 gives information about the water consumption in cotton processing. Cotton preparatory processes including bleaching consume nearly 38 % water out of the total water consumption for cotton processing (Karamkar 1999).

Fig. 2 Water usage in textile wet processing

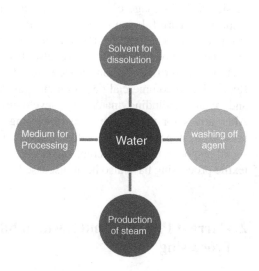

Process	Requirements (litres/kg of product)
Desizing	2.5–21
Scouring	20–45
Bleaching	2.5–25
Mercerization	17–32

Table 2 Water requirements for cotton processing (Menezes 2011)

Table 3 Temperature requirement of different textile chemical operations

Process	Cotton	Wool	Polyester
Scouring	100 °C (alkali boiling)	40 °C	70 °C
Bleaching	95 °C (peroxide)	80 °C (peroxide)	95 °C (peroxide)
Dyeing	85 °C (reactive dyes)	85 °C (acid dyes)	130 °C (disperse dyes)

2.2 Energy Consumption and Carbon Footprints

Energy is used in every process during the conversion of raw fibre into the finished product and its requirement depends upon the type and sequence of preparatory, dyeing, and finishing steps that are followed. Generally all thermal energy consumed in wet processing is for either heating the process bath or removing water from the textiles. Most of the textile operations are being carried out either in warm or high temperature conditions as shown in Table 3.

Energy requirements in textile processing also depend upon the type of machine used for processing. It is reported that long liquor ratio machines require more energy than short liquor ratio machines. The hank dyeing process using the cabinet machine requires more energy than the cheese dyeing process as the amount of liquor to be heated up is more. (Hasanbeigi 2010). Table 5 gives the general energy requirements for the processing of textile material. As this energy is mainly derived from the burning of fossil fuels it contributes to an increase in the carbon footprints of the textiles.

Globally the textile and clothing sector significantly contributes to the increase in carbon footprint (Muthu et al. 2012). Carbon footprints of natural fibres are comparatively lower than synthetic fibres as being derived from petroleum based raw materials the embodied energy of synthetic fibres is much more than natural fibres. On the contrary, cultivation of natural fibres such as cotton reduces the carbon footprints (Athalye 2012). Apparel and textiles account for approximately 10 % of the total carbon impact in the world. The estimated consumption of electricity for the annual global production of 60 billion kilograms of fabrics was estimated to be 1 trillion kilowatt hours (Zaffalon 2010) (Table 4).

2.3 Pollution Load and Waste Generation

(a) Pretreatment of Textile Materials

Pretreatment of textile material is carried out to remove the natural and added impurities to make the fibre accessible to dyes and finishing chemicals. In the desizing of cotton, sizes such as starch, polyvinyl alcohols (PVA), and the like are removed by the hydrolysis process. Presence of starch in the desizing effluent increases the biological oxygen demand (BOD). The desizing process is the main contributor to increased BOD in cotton processing. In general, about

Table 4 Energy requirements for the processing of textile material. (Hewson 1998)

Process step	Equipment used	Energy (kJ/kg)
Scouring and bleaching	Open width	3.0–7.0
	Beam	7.5–12.5
	Soft flow	3.5–16
Dyeing	Jigger	1.5–7.0
	Winch	6.0–17.0
Drying	Cylinders	2.5–4.5
	Stenter	2.5–7.5
Heat setting	Stenter	4.0–9.0

50 % of the water pollution is due to wastewater from desizing, which has a high BOD that renders it unusable (Ramesh Babu 2007). Noncellulosic substances including wax, pectin, proteins, and so on, present in the fibre wall are removed during the scouring process. Conventionally, scouring of cotton is carried out at temperatures reaching up to 120 °C in a strong alkaline medium. Auxiliary agents such as wetting agents, emulsifiers, and sequestering agents are also added to the scouring bath to improve its efficiency. The scouring process results in the effluent of high BOD, COD, TDS, and alkalinity (Naveed et al. 2006).

In the bleaching operation, natural and added coloured impurities are removed either by using reducing or oxidizing chemicals. Use of a chlorine based bleaching agent results in generation of adsorbable organo halides (AOX) (Wasif 2010). Hydrogen peroxide based bleaching is considered an ecological alternative to chlorine bleaching. However, this process also requires a large amount of water, high temperature, inorganic/organic stabilizers, and neutralizing agents. Auxiliary chemicals which are used in the bleaching bath such as phosphate based peroxide stabilizers increase the TOC and COD values of effluents. Upon neutralization of highly alkaline waste baths, large amounts of salts are produced. Apart from water and chemicals, the peroxide process also requires huge energy because it is carried out at the temperatures ranging from 90–120 °C. In wool processing, the carbonization process is done to remove the vegetable matter using sulphuric acid at high temperatures. The wool is steeped in the sulphuric acid solution, which causes the burrs to break up. This process also requires large amounts of water and chemicals for carbonization as well as neutralization of the sulphuric acid. In addition, the disposal of used sulphuric acid also poses a very big problem to the environment.

Alkylphenol ethoxylates (APEO) are used in the textile industry as wetting agents and detergents. Nonylphenol ethoxylates and octylphenol ethoxylates are important members in the family of APEO. They are also used in the scouring and bleaching operations. APEOs are found to be toxic to aquatic organisms due to their hormone-disruptive properties. They are highly persistent and nonbiodegradable. It creates problems in the wastewater treatment and discharge of treated water into the environment (Anon 2013).

(b) Colouration

Dyeing is the uniform application and fixation of colour to the textile material. Currently most of the dyeing is being carried out using synthetic dyes which are derived from the aromatic intermediates obtained from coal tar. Commercial dyes are not 100 % pure substances and the content of active ingredient varies from 20 to 80 %. They may contain heavy metals in their composition which will finally go into the product or into the effluent. Copper and chromium are mostly present in the metalized direct, reactive, and metal complex dyes. Potassium dichromate is used in the oxidation of vat and sulphur dyeing of cellulosic material. This results in Cr^{6+} in the effluent which is a known carcinogen. Metal salts are also used in the mordant dyeing of woolen goods. Presence of heavy metal in the finished product above prescribed limits is objectionable and banned by many eco standards. Removal of heavy metal atoms from the effluent is very difficult unless a very tedious procedure is followed. Wastewater composition from the dye house depends upon the nature of the dyeing technique and the machinery used. Apart from the dyes, many other auxiliary chemicals are used as additives in the dye bath. Carriers are used in the dyeing of polyester to facilitate its dyeing at atmospheric conditions. They are found to be toxic not only in the effluent but also are of concern to the health of workers. Table 5 lists various ingredients in the effluent of the respective dye class other than dyes.

Most of the dyeing auxiliaries are nonrecyclable and contribute to the high BOD/COD in the effluents. Usage of a huge amount of salt in reactive dyeing of cellulosic material in the exhaust method will result in effluent with high TDS (total dissolved solids). Removal of salt from the effluent needs to be done either by evaporation or by using reverse osmosis methods. The problem of highly coloured effluent containing unexhausted dye is a major issue in the dyeing of cellulosic fibres. It is approximately calculated that content of residual dyes in the exhausted dyeing of cotton materials is 10–50 % for reactive dyeing, 5–30 % for direct dyeing, 5–20 %, for vat dyeing, and 10–40 % for sulphur dyeing which goes into the effluent. Discharge of dye-containing effluents into the water bodies is undesirable, not only due to the colour, but also due to the release of toxic, carcinogenic, or mutagenic substances during their further degradation (Zaharia et al. 2009). The adverse effects of dyes on the aquatic environment can also be the result of toxic effects on the fish and other aquatic life forms due to their long persistence in the environment and buildup

Table 5 Various ingredients in the effluent of the different dye class

Dye class	Main ingredients
Direct, Reactive	Sodium chloride/Glauber's salt, surfactant, urea, and alkali
Vat, Sulphur	Sodium hydrosulphite, sodium sulphide, alkali, salt
Disperse	Carrier, acetic acid, sodium hydroxide, sodium hydrosulphite, dispersing agent
Acid	Glauber's salt, acetic acid/sulphuric acid, and surfactant

in deposits. Azo dyes derived from aromatic compounds show both acute and chronic toxicity (Börnick and Schmidt 2006). Reducing agents such as sodium hydrosulphite (hydros) and sodium sulphide used in vat and sulphur dyeing produce decomposition products which in turn increase the sulphide content in the effluent. Discharge of sulphides to drain is not permitted due to the deleterious effect of sulphuric acid formed by bacterial oxidation of liberated hydrogen sulphide.

(c) **Finishing**

Textile finishing is a very important operation in the wet processing sequence because it improves or modifies the hand and functionality of textiles. The finishing process can be classified into mechanical and chemical finishing. In chemical finishing, different kinds of chemicals are used to enhance the hand and functionality of textiles. Formaldehyde is one of the widely used chemicals in the manufacture of a large number of textile auxiliaries such as dye fixing agents, resin precondensates, softeners, and so on. Formaldehyde based substances are used as additives in many finishing formulations as cross-linking agents to improve durability. There is a probability of release of formaldehyde by these substances when they are put into use either during application or in the actual usage of the finished product. It is found that formaldehyde is a skin irritant and respiratory sensitiser. The International Agency for Research on Cancer (IARC) changed its classification of formaldehyde from a group 2A substance 'probably carcinogenic to humans' to a group 1 'carcinogenic to humans' in 2004.

Durable water repellency on textile material is being achieved with finishes that contain a polymer in which long-chain perfluoroalkyl groups are attached. These fluorinated polymers often contain residual raw materials and trace levels of long-chain perfluoroalkyl acids (PFAAs) as impurities. The residual raw materials and the products themselves may degrade in the environment to form long-chain PFAAs. Due to widespread use, long-chain PFAAs including perflurooctanoic acid (PFOA) and perflurooctane sulphonic acid (PFOS) have been detected globally in the environment. PFOA and PFOS, the most widely known and studied long-chain PFAAs, have been shown to be persistent in the environment, have a long elimination half-life in wildlife and humans, and have toxicological properties of concern (Lau et al. 2007).

Shrinkproof treatment of wool is normally done by pretreatment using chlorine followed by the resin treatment. Because in this process AOX generation in the effluent exceeds the permitted level (40 ppm), an eco-friendly alternative is required for antifelting finishing of wool materials.

Textile materials have been treated with various synthetic antimicrobial agents such as quaternary ammonium compounds (QAc), polyhexamethylene biguanide (PHMB), nano silver, and so on to impart antimicrobial efficacy. Among different antimicrobial agents used for the development of antimicrobial textile products, triclosan (2,4,4′-trichloro-2′-hydroxydiphenyl ether) has an important place due to its very low inhibition concentration (MIC) of less than 10 ppm for the bactericidal effects. However, when exposed to sunlight in the environment, triclosan breaks down into 2,8-dichlorodibenzo-p-dioxin which comes under toxic polychlorinated dioxins, hence its use is being restricted.

Polybrominated diphenyl ethers (PBDEs), a major class of brominated flame retardants are used for the textile finishing (Shin and Baek 2012). PBDEs are found to be very toxic to human beings (Darnerud et al. 2001). They are chemically similar to polychlorinated biphenyl (PCB), which has already been banned in many countries. Some flame retardants have toxicity due to the heavy metal content in their structure. Phosphorous, halogenated compounds, antimony, and zirconium based flame retardants increase toxicity of the wastewater. Dust containing flame retardants such as antimony oxide are also very toxic to the environment.

3 Approaches for Sustainable Textile Processing

To overcome the sustainability issues associated with textile processing, textile processing with eco-friendly chemicals with reduced water and energy usage can be the logical approach. Developments in sustainable processing can be classified as in Fig. 3.

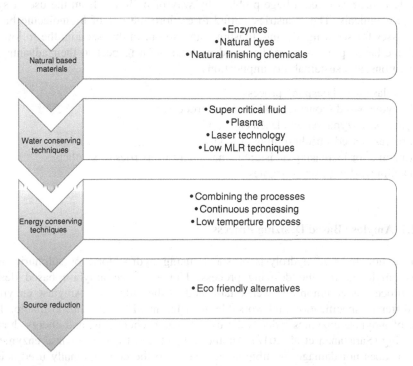

Fig. 3 Approaches for sustainable textile processing

The following sections discuss in detail the different developments taking place in the sustainable processing of textiles.

3.1 Enzymatic Techniques for Sustainable Textile Processing

The enzyme based processes are considered eco-friendly and sustainable. Among the different processes, the enzyme based textile processing methods are adopted by the industries effectively. The enzymes are biocatalysts and are specific towards the substrate. The handling of enzymes is safe and they are biodegradable. Enzymatic/biotechnological application in textiles can be considered as old as natural dyes. Indigo dyeing was carried out by our ancestors by the fermentation method which is the one of the most important methods of producing enzymes even today. The traditional fermentation methods that were used for textile processing are complex and time consuming. Hence with the introduction of synthetic chemicals during the eighteenth century, the biobased textile processing methods were not used. However, towards the end of the twentieth century, biotechnology got huge attention from researchers and several new techniques, so-called modern biotechnology developed. The researchers were able to develop effective biomolecules including enzymes through genetic engineering. At the same time, textile industries faced a huge problem by way of pollution from the use of synthetic chemicals. The industries started to explore the use of biomolecule based processes for scouring, dyeing, finishing, and so on. In this section, the following enzyme based processes are discussed in detail with respect to their advantages, limitations, and sustainability implications.

- Amylase based desizing process
- Enzyme based scouring and bleaching processes
- Catalase enzyme based bleach cleanup
- Protease based shrinkproofing of wool
- Enzyme for increasing moisture management properties of synthetics
- Denim washing using enzymes

3.1.1 Amylase Based Desizing Process

Conventionally acid hydrolysis of starch using hydrochloric or sulphuric acid was employed for the desizing process. However, the amylase based desizing process is common and well established in the industries. Amylase enzymes of different specification and workable at different pH ranges are available. The use of amylase enzymes reduces the use of water, chemicals, and energy during desizing (Saravanan et al. 2012). Another advantage of using amylase enzyme is that it does not damage the fibre as compared to the conventionally used acids. Irrespective of the enzyme or acids, the starch is converted into sugar and its derivatives during desizing. Though sugar is nontoxic, it is not recyclable and the effluents create many environmental problems. Hence, the enzyme based desizing process needs to be modified to address the above problem. In this direction, attempts have been made to desize using amyloglucosidase enzyme instead of a conventional amylase enzyme in order to hydrolyse starch into single glucose units. After that, the glucose oxidase enzyme was used to generate hydrogen peroxide from the glucose generated during desizing (Eren et al. 2009).

3.1.2 Enzyme Based Scouring Process for Cotton

Conventionally, cotton textile is treated with sodium hydroxide for scouring purposes at high temperature and pressure. An alkaline pectinase enzyme based process was developed as an alternative to alkaline scouring which can also be done at moderate temperature. The enzyme gently acts on cotton fabric and leads to less damage, less pilling, and uniform dye uptake compared to conventional alkaline scouring. Apart from pectinase enzymes, other enzymes such as cellulase, protease, cutinase, and others have been proposed by the researchers for the scouring of cotton. The enzyme is substrate (pectin) specific in nature and is not able to remove noncellulosic impurities completely due to poor access to the substrates when used on industrial scale. In order to make the enzyme based scouring process industrially adoptable, it is proposed to use a combination of enzymes such as pectinase + cellulase, pectinase + protease, pectinase + cutinase (Agarwal et al. 2008), pectinase + lipase (Kalantzi et al. 2010), or pectinase + xylanase (Battan et al. 2012). Attempts have also been made to combine enzyme based scouring and activator assisted bleaching in single bath to remove the noncellulosic impurities from the cotton textiles (Hebeish et al. 2009).

3.1.3 Enzyme Based Shrinkproofing for Wool

It is well documented that papain, a proteolytic enzyme of plant origin was used for shrinkproofing of wool in earlier years. However, the wool had to be pretreated with a reducing agent like sodium sulphite to achieve an effective shrink-resistant property before enzyme treatment. The efficiency of the proteolytic enzyme papain in conferring shrink resistance to wool tops and woven fabrics has been enhanced by pretreatment of the wool with lipase (El-Syed et al. 2001). The proteolytic enzymes from bacterial sources are commonly used for imparting shrink resistance and handle to wool. This enzyme is applied on wool in alkaline pH. A pretreatment with alkaline peroxide before the enzyme treatment is necessary for effective activity of the enzyme on wool. The enzyme activity on wool depends on the pH, temperature, time of the treatment, and concentration of enzyme.

An acid protease enzyme from fungi source was applied on wool serge fabrics to improve the handle and shrink resistance and compared with the conventional alkaline protease. The results indicated that acid protease enzyme improved the handle as much as the alkaline protease but not shrink resistance (Raja and Thilagavathi 2010). Cortez et al. (2007) investigated the transglutaminase mediated grafting of silk protein on wool and its effects on wool properties. The grafting leads to significant effect on the properties of wool yarn and fabric, resulting in increased bursting strength as well as reduced values of felting shrinkage and improved fabric softness.

3.1.4 Enzyme Treatment to Increase Dye Uptake

Enzyme pretreatment on wool fabric decreased the resistance of the fibre to dye diffusion and so it increased the adsorption rate constant and decreased the apparent activation energy for the dyestuff when compared to untreated fabric. Enzyme pretreatment of wool fabric with trypsin increased exhaustion of natural dyes such as crocin, beta-carotene, curcumin, chlorophyll, and carmine without change in fastness property (Liakopoulou et al. 1998).

3.1.5 Catalase Enzyme Based Bleach Cleanup

Currently, hydrogen peroxide is the preferred bleaching agent for the bleaching of textiles. When hydrogen peroxide bleaching of cotton is done prior to dyeing of cotton with reactive dyes, the presence of residual peroxide will create uneven dyeing and promote hydrolysis of the reactive dye. In order to prevent the above process, normally fabrics are treated with chemicals to completely clean up the peroxide before dyeing with reactive dye. Conventionally, the hydrogen peroxide bleached fabrics have to undergo two to three washes with water and treatment with sodium thiosulphate for removing the peroxide. The process requires a higher amount of water, energy, and time. Biotechnology has come to the rescue to reduce the washes with the introduction of catalase enzymes for removing the residual hydrogen peroxide. Catalase enzyme can effectively remove the residual hydrogen peroxide in place of chemicals. The catalase based processing requires less water in terms of reduced rinsing steps and it is also eco-friendly. The catalase treated fabrics exhibit uniform dye uptake and good colour values (Amorim et al. 2002). Efforts have been made to reuse the catalase treated bleach cleanup bath for subsequent dyeing also. (Tzanov et al. 2001).

3.1.6 Enzyme for Increasing Moisture Management Properties of Synthetic Fabrics

The use of synthetic fibres such as nylon, polyester, and acrylic for the production of apparel and technical applications has a huge share of the textile market. The limitation of these fibres in the apparel sector is their inherently poor moisture management properties. Enzymes are traditionally considered as suitable for application on natural fibres. Nylon fabrics are treated with the protease and mixture of protease and cutinase enzymes to improve moisture uptake. The above enzymes hydrolyse the surface amides present in the nylon to improve moisture uptake (El-bendary et al. 2012). Similarly, polyester based textile materials were treated with esterase, lipase, and cutinase enzymes to improve the moisture absorption characteristics as well as to improve the surface softness and reduce pilling. Conventionally, the polyester fabric has to be treated with sodium hydroxide to improve the above properties. However, such treatment led to higher strength loss and damage to the fabric. The use of lipase and cutinase enzyme substitutes the harsh chemical treatment and the process is considered eco-friendly (Heumann et al. 2006). Acrylic fibres are modified by the nitrilase enzyme (Matamá et al. 2007). The enzymatic modification processes for synthetic fibres are still at a nascent stage and require further studies for commercialization.

3.1.7 Denim Washing Using Enzymes

Nowadays denim is accepted by all age groups and genders as a popular form of clothing. As per the Global Market Report 2012, global denim fabric production capacity is over 7 billion yards and is estimated to increase to 9 billion yards by 2021. The use of cellulase enzyme for denim washing is one of the established and successful applications of biotechnology in textiles. About 80 % of the denim washing industries use cellulase alone or in combination with pumice stone for processing. One of the problems associated with the use of cellulase for denim washing is backstaining of denim due to redeposition of indigo dye (Pazarlioğlu et al. 2005). The backstaining problem due to the action of cellulase enzyme can be reduced with subsequent laccase enzyme treatment (Tarhan and Sarıışık 2009). The laccase enzyme treatment alone or in combination with cellulase provides optimum washing of denim with less backstaining (Montazer and Maryan 2010).

3.2 Use of Natural Eco-friendly Materials for Processing

3.2.1 Biosurfactants

Surfactants derived from microorganisms are called biosurfactants Microorganisms produce biosurfactants/antibiotics during the stationary growth phase when the nutrients are exhausted in the medium. This is one of the mechanisms of population control in microorganisms especially in an environment deficient in nutrition and space. The nature of antibiotics produced by microorganisms depends on the type of strain and the medium used for growth. The group of antibiotics produced by microorganisms are glycolipids, phospho-lipids, ornithine-lipids, fatty acids, lipoproteins and lipopeptides, and aminoacid lipids. The antibiotics produced by microorganisms also have surfactant properties (Desai and Banat 1997). Hence, the application of these substances in textile material will have a dual function such as surfactant and antimicrobial properties apart from their nontoxic nature to humans and the environment. Lipopeptides are a group of antibiotics produced by *Bacillus* sp. Lipopeptides have high antimicrobial activity and also have surfactant property. Lipopeptides are resistant to harsh environments such as wide ranges of temperature, pH, salt concentration, acids, and so on (Makovitzki et al. 2006; Mageshwaran et al. 2012). Savarino et al. (2009) derived biosurfactants from the urban waste biomass and used them for the dyeing of cellulose acetate material. Dyeing results indicated that there is no significant difference between the surfactant obtained from the biomass and synthetic surfactants in terms of surface activity and other properties.

3.2.2 Natural Dyes

The art of dyeing is as old as our civilization. From ancient times to the nineteenth century, natural dyes were the only source to colour textiles. Natural dyes are the colouring materials obtained from natural resources of plant, animal, mineral, and microbial origins. The use of natural dyes started diminishing after the invention of synthetic dyes in the later half of the nineteenth century. In recent years, the use of natural dyes is gaining renewed interest mainly among eco-savvy people. Natural dyes are considered eco-friendly, skin friendly, and are found to have health benefits to the wearer. Natural dyes can be used for dyeing almost all types of natural fibres and recent research shows that it can be used to dye some synthetic fibres also. Apart from the textile application, natural dyes are also used for different applications including food colour, medicine in traditional medical systems, and leather processing, among others. There are several challenges and limitations associated with the use of natural dyes.

The availability of natural dyes needs to be increased in a sustainable manner by utilizing the byproducts and wastes from agriculture and agroprocessing industries and judicious collection of forest produce. This may be supplemented by growing important dye-bearing plants on wastelands and marginal lands which may also provide an alternative cash crop to cultivators. Establishment of proper characterization and certification protocols for natural dyes would definitely improve consumer confidence in natural dyed textiles and would benefit both producers and users. If natural dye availability can be increased by the above-described measures and the cost of purified dyes can be brought down with a proper certification mechanism, there is a huge scope for adoption of these dyes by small-scale dyeing units as they lack the resources to install and operate expensive effluent treatment plants needed to bring the synthetic dye effluent within the limits set by regulatory authorities. If at any time in the future, the availability of natural dyes can be increased to very high levels by biotechnological interventions such as tissue culture or genetic engineering resulting in mass production of these dyes by microbes at low cost, only then can their usage become sustainable for mainstream textile processing. At the level where scientific developments stand today, natural dyes are a sustainable option only for small-scale applications and they can complement synthetic dyes as an eco-friendly option for the environment-conscious consumer and a means of providing a livelihood to various stakeholders of the natural dye value chain.

3.2.3 Natural Product Based Finishing

The recent improvement in living standards has increased awareness about the performance of textiles which has led to the development of functional textiles. Textile materials are treated with different chemicals to impart various functionalities. These chemicals may be antimicrobial, medicinal, UV protective, mosquito repellent, and so on, to impart a specific functionality. Natural products due to their eco-friendly and skin-friendly properties play an important role in the sustainable functional finishing of textiles (Joshi et al. 2009). The natural product based functional finishing of textiles can be broadly classified into the following areas.

- Antimicrobial finishing
- Aroma/deodorant finishing
- UV protective and flame retardant finishing

(a) **Antimicrobial Finishing**

Textile materials are good media for the generation and spreading of microorganisms. Keratin and cellulose can be the nutrients for the growth of microorganisms (Purwar and Joshi 2004). The growth of microorganisms in textile materials causes unpleasant smells, staining, loss of mechanical strength, and the like, and it can create health-related problems for the wearer. Hence, antimicrobial finishing of textile materials is necessary to protect the wearer from harmful microorganisms. Antimicrobial agents were used on textiles thousands of years ago, when ancient Egyptians used spices and herbs as preservatives in mummy wraps (Seong et al. 1999). An antimicrobial finish is a method to reduce the spread of microorganisms by either killing or inhibiting their growth through contact with the fabric surface (Huang and Leonas 2000).

There are several studies in the literature on natural dyes extracted from Quercus infectoria, curcumin, and so on (Han and Yang 2005; Singh et al. 2005) which have been used to give both dyeing and antimicrobial finishing to textile materials. The tannins present in plants are a group of water-soluble polyphenols in the molecular weight range of 300–5000 Da. The tannins can be divided into two groups, namely, hydrolysable tannins and condensed tannins. The hydrolysable tannins are usually compounds containing a central core of glucose or other polyhydric alcohols esterified with gallic acid (gallotannins) or hexahydroxydiphenic acid (ellagitannins). The condensed tannins are polymers of flavan-3-ol (catechin) units. They are also called proantho-cyanidin. It is well documented that the tannins present in the different parts of the plants such as bark, leaf, fruit, and so on have antimicrobial properties to several strains of bacteria through in vitro studies (Chuang and Wu 2007; Han et al. 2007; Min et al. 2008). The mechanism of antimicrobial actions of tannins can be summarized as follows: (i) tannin binds with the proteins and enzymes present in the cell wall of microorganisms and inhibits their growth; (ii) tannins also have the ability to bind with vital metal ions used by the microorganisms for their growth (Biradar et al. 2008); (iii) The gallic acid released from tannins is able to inhibit the growth of microorganisms by reacting with their cell wall and inhibiting their metabolism. The antimicrobial efficacy of natural plant extracts depends on the source of tannins and their concentration on the substrate. Generally, the natural tannin-treated materials exhibit good antimicrobial efficacy against gram-positive bacteria rather than gram-negative bacteria. In order to produce antimicrobial efficacy against gram-negative bacteria, a higher concentration of natural materials on the substrate is required. However, the absorption of natural substances on textile material cannot be increased after a particular saturation point and that is a drawback. The following are the most studied natural products for imparting antimicrobial property to the textile materials.

(i) **Chitosan**

Chitosan is a derivative of chitin, which is the second most abundant natural polymer. Its structure is very similar to that of cellulose except an amino group replaces one of the hydroxyl groups. Chitosan can destroy bacteria by converting its amino group into an ammonium salt in dilute acid solutions. Quaternary ammonium salt of chitosan can destroy the cell wall of the microorganism by connecting to its negatively charged protoplasm (Kim et al. 1998). Lee et al. (1999) evaluated chitosan as an antibacterial agent along with a blood repellent finish. Chitosan was applied to wool as a shrinkproofing polymer. Owing to the hydrophobic nature of wool, treatment with chitosan required pretreatments so that polymers can adhere to the surface. It is reported that wool was oxidized with potassium permanganate prior to the application of chitosan. Although chitosan had conferred durable antimicrobial ability to wool, the handle of the treated wool fabric was adversely affected (Heish et al. 2004). Wool fabric was finished with chitosan along with henna natural dye for dyeing and antimicrobial finishing (Dev et al. 2009).

(ii) **Neem Extract**

Azadirachtin, a tetranortriterpenoid with molecular formula $C_{35}H_{44}O_{16}$ has been identified as an active substance of neem. It is an insect anti-feedent and ecdysis inhibitor. It contains a large number of functional groups and is sensitive to acids, bases, and UV light. The cotton fabric treated with the neem extract shows very good resistance to both gram-positive and -negative bacteria but the durability is found to be very poor. One of the major limitations with the use of natural products for antimicrobial finish is lack of durability of the finish. Most of the applied antimicrobial agents can be removed during washing because they do not have any affinity to textiles or they are not fixed on the textiles. Microencapsulation is one method used to trap the active antimicrobial agent using wall materials such as modified starch, gum acacia, sodium alginate, and so on and then applied on the textiles (Thilagavathi and Kannaian 2010).

(b) **Fragrance and Deodorant Finishing**
Fragrance-infused textile materials are now marketed by various companies and have wide acceptability among consumers. These kinds of fabrics are also called aromatherapy textiles because apart from giving a freshness feeling to the wearer they also give medicinal value such as relieving stress, cough, allaying fear, and imparting an antibacterial property. Generally, aromatic chemical fragrance compounds and essential oils are used for producing the scented materials (Achwal 2004).

(c) **UV Protective and Flame Retardant Finishing**
Ultraviolet rays (UVA and UVB) represent a very low fraction in the solar spectrum (200–400 nm) but affect all living organisms and their metabolisms due to their higher energy. These radiations can cause a range of effects from simple tanning to skin cancers, if the skin is unprotected (Sarvanan 2007). Many reports are available for the UV protective finishing of textile material using naturally derived materials. It was reported that cotton fabric treated with plant extracts derived from the dried fruit of harda (Terminalia belrica) and rinds of pomegranate (Punica granatum) resulted in very good UV protection 50+ (UPF). UV protection functionality was retained by the fabric even after 10 cycles of washing (ISO 105 C10-2006 Test No 8 B No 2). Colouration produced by this process also had good fastness with washing (ICAR-CIRCOT 2014–15). Basak et al. (2015) reported a finishing process for cotton textiles with flame retardant property using banana pseudo-stem sap. The extracted sap was applied to the premordanted bleached and mercerized cotton fabrics using different pH conditions.

3.3 Promising Waterless Processing Technologies

Several waterless textile processing technologies have been developed by researchers. The following technologies have been developed up to the commercial level.

- Supercritical fluid processing
- Plasma processing
- Laser technology

3.3.1 Supercritical Fluid Processing

A substance at a temperature and pressure above its critical point is a supercritical fluid, where distinct liquid and gas phases do not exist. Supercritical fluids have properties between liquids and gases. Though many substances can be converted into supercritical fluids, carbon dioxide has emerged as a suitable substance as it is cheap, nonflammable, nontoxic, and its critical point ($T_c = 31.4 \,°C$, $P_c = 73.7$ bar, Fig. 4) is much lower than that of many other substances. It has been employed for extraction of natural substances and dry-cleaning purposes. The concept of using supercritical carbon dioxide (SC CO_2) for processing of textiles was first established by Professor Schollmeyor of Germany in the late 1980s and received attention for practical dyeing applications since the last decade of the twentieth century (Banchero 2013).

As CO_2 is a nonpolar molecule, it behaves as a nonpolar organic solvent in supercritical state. Therefore water insoluble dyes such as disperse dyes can be easily dissolved in SC CO_2 without the use of dispersing agents. A cosolvent or a modifier may be added to improve the solubility of slightly polar solutes. Materials dyeable with such dyes including polyester, polypropylene, and polylactic acid, and so on, which pose problems in dyeing under aqueous systems can be easily dyed using SC CO_2.

Sustainability Benefits

Following are the sustainability advantages of using supercritical carbon dioxide processing over conventional aqueous based processing.

- Unused dye can be easily separated from carbon dioxide after completion of the dyeing process by bringing the dye solution to atmospheric conditions. Dye powder is separated and carbon dioxide also recovered. Recovered dye powder and CO_2 can be reused. This will not generate any effluent in the dyeing process. Problems and costs related with the effluent treatment become null.
- As the processed material is not carrying any solvents, drying is not required which saves the cost of energy and capital cost of drying machines.

Fig. 4 Phase diagram of carbon dioxide (Gebert et al. 1994)

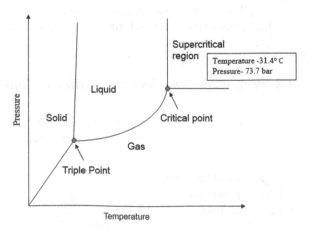

- Dyeing time is reduced due to the higher rate of dyeing. Therefore the rate of production is comparatively higher.
- The nonpolar nature of CO_2 results in good solubility of dispersed dyes. Thus dispersing agents are not needed.
- Though the cost of machinery is high, operational cost is nearly 50 % lower than the aqueous processing due to higher rate of dyeing (Vander krann 2005).

Dyeing Process

As SC CO_2 processing systems operate at high temperature and pressure, the machine essentially consists of an HTHP vessel, CO_2 tank, compression pump, container to keep dyestuff poder, and a circulation pump. First, the textile material to be dyed is wound on the bcam and kept in the HTHP vessel. Then the system is made to attain the required pressure and temperature and CO_2 is brought into supercritical state. It dissolves the dyc in this state and the resultant dye liquor is circulated in and out of the HTHP vessel by a circulating pump. After completion of the dyeing cycle, dye solution is depressurized to separate dyes and CO_2 gas. Clean SC CO_2 is then circulated to remove the unfixed dyes from the dyed material. Thus water is completely avoided and there is no effluent generation.

Polyester (PET) dyeing in SC CO_2 is the most researched and has come up to the commercial level. The experimental range investigated for the PET processing is 60–150 °C and 100–350 bar pressure (Van der Kraan 2005). It is recommended to heat set the textiles prior to dyeing to avoid strength reduction and shrinkage. Polypropylene (PP) fibres give a lot of trouble in aqueous dyeing due to their high crystallinity and nonpolar nature. These can be dyed with good all-round fastness properties by using hydrophobic long alkyl chain dyes under SC CO_2 (Miyaski et al. 2012). This would increase the market potential of PP in different sectors. Polylactic acid (PLA) fibre is a sustainable synthetic fibre due to its renewable raw material origin. But it application in the textile industry is limited due to its poor resistance to aqueous based processing which results in high shrinkage and loss in mechanical properties. However, studies conducted on dyeing of PLA using disperse dyes in SC CO_2 medium showed that the dyed material retained many of its mechanical properties (Wen and Dai 2007).

Dyeing of Natural Fibre Textiles in SC CO_2

Much success has not been achieved in the dyeing of natural fibre textiles in SC CO_2 due to poor solubility of polar dyes in SC CO_2 and poor affinity of nonpolar disperse dyes towards natural fibres. Also, lack of interaction between natural fibre polymers and SC CO_2 molecules results in very less swelling of these fibres leading to low dye uptake (Bach et al. 2002). Different approaches such as fibre modification, dye modification, or process modification by adding other solvents with CO_2 have been tried to uncover a solution for this problem. Silk and wool were successfully dyed in SC CO_2 with nonpolar dyes containing vinylsulphone or dichlorotriazine reactive groups with good fastness properties. The dyeing of natural fibres from a reverse-micellar system in supercritical carbon dioxide was attempted using ammonium carboxylate perfluoropolyether as surfactant. Higher colour depths were obtained in silk when both SC CO_2 and the textiles were saturated with water. In the case of cotton SC CO_2 dyed material did not have good fastness properties (Sawada et al. 2003). Some efforts to use this technology in wax removal/scouring of cotton textiles have been reported (Beck and Lynn 1997). Supercritical carbon dioxide assisted silicone based finishing of cotton fabric was also attempted recently (Mohamed et al. 2013).

Present Status of Commercialization

DyeCoo Textile systems, a Dutch Company, was the first one to launch a commercial production system in 2007 for SC CO_2 dyeing of scoured polyester in batches of 100–150 kg. Dyeing of natural cellulosic fibres such as cotton by using SC CO_2 has not yet been successful at the commercial level (Dyecoo 2010). This is the major stumbling block in the commercialization of this technology because of the good share of cotton and PET/cotton textiles in the woven and knitted sector. DyeCoo is working with partners to develop and deliver dye and chemical products to support the waterless dyeing process to obtain the high level of colour fastness and performance and expects to come up with modified reactive dye systems which can be used for dyeing cellulosics using SC CO_2 medium.

3.3.2 Plasma Technology for Textile Processing

In 1929, Lewi Tonks and Langmuir used the terminology 'plasma' to describe a collection of charged particles. Plasma is called the fourth state of matter. If a substance in its gaseous state absorbs higher energy, the outermost electrons which are present in the atoms will escape from the nucleus' control and become free electrons. The atom becomes positively charged due to the loss of electrons. This chemical state of a substance is called plasma. It consists of positive, negative ions, atoms, electrons, molecules, radicals, and photons. As the plasma is highly reactive in nature, it has the ability to react with other substances leading to various chemical fusions and fissions. These effects can alter the surface structure of textiles and resulting different functionality.

Sustainable advantages of plasma process over the conventional process (Shishoo 2007) are:

- No or much less water (cooling) is used in plasma technology. Therefore no effluent is generated in this process. This process is completely dry technology.
- Chemical and energy consumption in the process is comparatively very low.
- No extra drying process is required.
- Because the surface structure of materials is only modified by plasma, the other properties of textiles will mostly not be affected.
- Productivity speed in plasma technology is higher than current wet processes.
- A small amount of plasma is enough to produce the required effect and different kinds of gaseous chemicals can be processed in the same equipment.
- High innovation potential which would lead to development of new products.

Plasma Reactors

Plasma can be generated by using different types of power supply and by using different frequency of electromagnetic radiations. This is classified as follows.

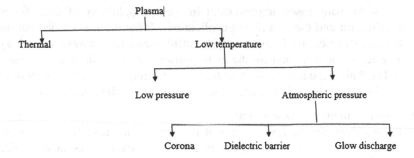

Fig. 5 Different types of plasma

Low frequency: 50–450 kHz
Radio frequency: 13.56 or 27.12 MHz
Microwave: 915 MHz or 2.45 GHz

Different types of plasma are given in the flow diagram (Fig. 5). In as much as textile materials are heat sensitive, low temperature nonthermal plasma is very useful for textile modification.

Low pressure plasma requires special conditions to produce plasma and was developed many years ago. Developing low pressure conditions in the bigger size commercial reactors is very costly and also this low pressure plasma technology is not highly suitable for integration into the current textile processing line of machinery. Therefore for practical textile applications, atmospheric pressure plasma (APP) reactors are found to be highly suitable due to their capability for inline and online application and lower cost. Plasma technology has vast potential to contribute towards sustainable textile processing due to very minimum environmental impact. This technology is contributing to sustainable processing in different ways. In some cases it completely replaces the wet processing and in other cases it reduces the time and temperature of processing which results in a huge saving of energy and harmful chemical usage. The following section discusses in detail the application of plasma technology for various textile processes.

(a) **Desizing and Scouring Using Plasma Technology**

It is reported that desizing of polyvinyl alcohol (PVA) sized cotton fabric can be done only by using cold water washing after plasma treatment using air + helium + oxygen plasma. It was found that plasma treatment using helium + oxygen plasma could be an effective alternative to conventional enzymatic desizing of starch based sizes (Cai et al. 2003 and Kan et al. 2014). Reports are available for the reduction of cotton scouring process time using RF plasma excited with oxygen as the plasma gas (Sun and Stylios 2004). This will offer the advantage of saving energy and washwater in cotton processing.

Similarly in degumming process of silk, washing temperature and concentration of soap can be reduced significantly using plasma pretreatments (Long et al. 2008). Hydrophobicity of synthetic fibres such as polyester, nylon, and polypropylene is well known due to the absence of polar groups in these fibres. In many cases, improvement in the hydrophilicity of these fibres is required for end use in different applications. Presently it is carried out using the wet chemical method by using harmful chemicals. Plasma technology is an effective tool to increase the hydrophilicity of synthetic fibers (Mehmood 2014). This treatment not only reduces the wetting time of the substrates but also increases the antistatic and adhesion property of these materials.

(b) **Improvements in Colouration**

Plasma treatment of textile material is found to increase the percentage of exhaustion of dyes which results in lesser amount of dyes in the effluent. This also reduces the dyeing time and temperature. This will be very useful in the effective usage of dyes which are giving a lot of trouble in the effluent treatment. Though this is applicable to most of the textile fibres theoretically, promising results are obtained for protein fibres especially for wool (Kan et al. 2006). Pigments are used in the printing of cotton and its blended textiles. However, there are some problems in pigment dyeing of cotton materials including lower colour yield and poor rubbing fastness. It is reported that oxygen plasma pretreatment of cotton fabric showed a positive influence on increasing colour yield, levelness, and improving the colour fastness to crocking of pigment (red pigment) dyeing to cotton fabric (Man et al. 2014). This will encourage the application of pigments on cotton materials.

(c) **Plasma Based Finishing Processes**

Plasma technology is mostly exploited in the finishing of textile materials. Enormous possibilities are available to develop innovative textile products using this technology. Plasma polymerization on cellulosic, polyester, and nylon materials using fluorine based gases results in good water and oil repellent functionality by reducing the surface energy of these substrates (Li and Jinjin 2007; Samanta et al. 2010; David et al. 2013). This is a completely dry process and does not require any further drying and curing processes. This will save a lot of water, energy, and chemicals. The main advantage of hydrophobic finishing of textiles using plasma technology is that comfort properties such as water vapour permeability, air permeability, and so on are not affected which will be very useful for textiles for apparel purposes. An eco-friendly alternative is required for antifelting finishing of wool materials. This can be achieved by pretreating the woolen textiles using the air/oxygen plasma followed by coating with polyurethane based resin. The advantage of this plasma based pretreatment is elimination of the chlorine based wet treatment and also avoiding the use of non eco-friendly chlorine based chemicals (Hartwig 2002).

(d) **Commercial Machinery**

Many European and American plasma equipment suppliers understood the practical application of plasma technology for textile materials and they have come out with a new kind of plasma based machinery for textile materials. Table 6 gives a few of these suppliers.

Table 6 Commercially available plasma reactors for different applications

Manufacturer	Application
Dow Corning, USA	Surface modification and coating (Dow corning corporation 2007)
Sigma, USA	Surface modification and coating (Sigma technologies 2006)
Apjet, USA	Water and stain repellency (www.apjet.com)
Acxys, France	Wettability, water repellent (www.acxys.com)
Diener, Germany	Cleaning, etching, activation, polymerization (Diener 2006)
Plasmatreat, USA	Self-cleaning, flame retardancy (www.plasmatrcat.com)
Arioli, Italy	Water repellent (www.arioli.biz)
Vito, Belgium	Cleaning, activating, coating (www.vitoplasma.com)
Softal, Germany	Water repellent, wettability (www.softol.de)
Europlasma, Belgium	Surface coating, Water repellent (www.europlasma.be)

3.3.3 Laser Technology

Laser technology has the ability to achieve a faded look and worn-out effect on denim materials. The use of laser engraving with the aid of computer design gives special printing effects. A variety of colour removal with little or no damage to the other properties of denim material can be achieved by using different laser parameters. Laser based finishing technology has an edge over other conventional processing techniques due to its waterless nature, accuracy, and possibility of innovation (Dascalu et al. 2000; Ondogan et al. 2005; Kan 2015). The Spanish Company, Jeanologia, who produce a denim garment processing machine using laser technology, claim that water consumption can be reduced dramatically by using their technology.

4 Developments in Dyeing

4.1 Salt Reduction in Reactive Dyeing

Reactive dyes are water soluble anionic dyes which differ from other classes of dyes by undergoing chemical reaction with fibre polymer. They differ among themselves in terms of nature and number of reactive groups. Major issues with the reactive dyes for the dyeing of cellulosic material are their instability in aqueous condition and requirement of higher ionic strength in the dye bath which require a huge quantity of electrolytes. Dyeing 1 kg of cotton with reactive dyes requires an average of 0.6 kg NaCl and 40 g reactive dye. Many developments have taken place to reduce the salt requirement and to increase the dye affinity in reactive dyeing. The research in salt-free dyeing can be broadly classified into two áreas:

- Use of organic salts
- Cationization of cotton

(a) **Use of Organic Salts**

   In this area of research, organic salts are used in place of inorganic slats such as sodium chloride and sodium sulphate. Prabhu and Sundarajan (2002) proposed sodium citrate as an alternative to inorganic electrolytes. Bleached cotton fabrics were dyed with reactive and direct dyes with the use of sodium chloride and sodium citrate. The results showed that fabric dyed satisfactorily with the use of sodium citrate and there was significant reduction in total dissolved solids. Khatri et al. (2012) used trisodium nitrilotriacetate as an alternative to sodium chloride for dyeing of cotton in the pad-steam dyeing method. The result showed that the above organic salt can be used to dye cotton satisfactorily with reactive dyes with minimum pollution load. Ahmed (2005) proposed the use of sodium edate as an alternative for salt. Guan et al. (2007) also used polycarboxylic acid sodium salt for salt-free dyeing. Yu et al. (2014) have reported that sodium oxalate can be used as an organic electrolyte for cotton to replace the inorganic electrolytes.

(b) **Cationization of Cotton**

   Several studies in the field of salt-free reactive dyeing used cationic agents such as quaternary ammonium based chemicals for imparting a positive charge on cotton fabrics. The agents are organic in nature and considered to produce a minimum pollution load. The cationic agents used are 1-amino-2-hydroxy-3-trimethylammoniumpropane chloride (Wang and Lewis 2002), 3-chloro-2-hydroxypropyl trimethyl ammonium chloride–CHTAC along with 40 gpl sodium hydroxide (Hashem 2007). Several studies were conducted to use cationic polymers instead of cationic agents to impart a cationic nature to cotton. Generally, the process employed for coating polymers is the pad-dry process. The polymers used are dimethylamino ethylmethacrcrylate (Fatma and El-Alfy 2013), polyepichlorohydrin dimethylamine (Wu and Chen 1993), polyamide-epichloro-hydrin resin (Burkinshaw et al. 1989), poly(4-vinylpyridine) quaternary ammonium compound 1 % owm (Blackburn and Burkinshaw 2003), dentrimers (Burkinshaw et al. 2000), and amino terminated polymers (Zang et al. 2007).

Chitosan along with assisting agents for penetration and starch derivatives have been used through pretreatment before dyeing with reactive dyes without salt. Chitosan and its derivative O-acrylamido- methyl-N-[(2-hydroxy-3-trimethylammonium) propyl] chitosan chloride (NMA- HTCC) was used for the dyeing of cotton (Gentile 2009; Lim and Hudson 2004). Wei at al. (2012) used cationic starch and its hydrolysed starch in salt-free dyeing of reactive dyes.

4.2 Best Available Method of Reactive Dyeing

The cold pad-batch dyeing method of cotton textiles is considered the best available technology for the dyeing of cotton with reactive dyes. This is a semicontinuous method of reactive dyeing of cotton and woolen materials. Ready for dyeing fabric is padded with liquor containing premixed reactive dyes and alkali (sodium silicate or sodium carbonate). The fabric is then batched onto rolls and covered with polythene sheets to prevent evaporation of water and stored for 6–12 h. After the batching period, the material is washed with water and hydrolysed dyes are removed by soaping. Following are the sustainable advantages this method.

- In this method, salts, lubricants, leveling agents, fixatives, and defoamers are not used. Hence, the effluent load is much less.
- Because the dyeing and fixation are done at room temperature energy consumption is comparatively much less than other methods.
- Another advantage of this method includes savings on water and labor. Water consumption for pad-batch dyeing with beam washoff is only 10 % of the amount used compared to the dyeing of fabrics using winch machines (90 % reduction; Marbek Resource Consultants 2001).

4.3 Development in Commercial Dyes

(a) **Sulphur and Vat Dyes**

Archroma (formerly known as Clariant) has developed a process called 'advanced denim' using prereduced liquid sulphur dyes and sugar based reducing agents. It is claimed that this technology uses less energy, water, and has a lower pollution load. DyStar patented the DyStar indigo vat 40 % solution, which is prereduced indigo liquid. It is claimed that denim production will be cleaner with a reduction in sodium hydrosulfite usage by using this product.

(b) **Reactive Dyes**

Many commercial reactive dyes are available which require less water and energy for dyeing. Avitera® SE which is a multifunctional reactive dye (three groups) was introduced by Huntsman at ITMA 2011. These reactive dyes are mainly used for the dyeing of cellulosic fibres, and are claimed to save energy and time with reduced water consumption. Eriopon LT, a special auxiliary, is being used to assist the washing-off of unfixed dye from the fabric. DyStar developed Remazol Ultra® RGB reactive dye range for the problematic deep shades on cellulosic fibres. It is claimed that this range gives increased productivity and reduces the effluent load due to preventing redyeing. DyStar's Levafix® CA dyes are of interest as they are AOX free. It is claimed that this range of dyes gives high fixation with very good light fastness properties (Lewis 2014).

New reactive dyes (e.g., Lanasol® CE, Ciba) and optimized ranges of metal-free acid dyes (Sandolan® MF, Clariant) have been developed to achieve a balance of economy and performance comparable to chrome dyes in targeted applications such as piece and hank dyeing, especially bright fashion shades for woolen materials. Basolan® AS (BASF) inhibits loss of bulk in package dyeing and damage to the dyed wool. This type of technology is finding use in a number of applications, particularly in dyeing of wool–polyester blends at temperatures up to 120 °C where the protective effect is claimed to be better than conventional formaldehyde-release agents (Cookson et al. 1995).

5 Digital Printing for Cleaner Production

Textile printing is one of the most useful techniques to produce fashion textiles. Conventionally automatic rotary screen printing technique is mostly used for the textile printing. Digital printing is one of the most modern achievements in the printing field of the textile industry. It is described as an inkjet based method of printing of colourants onto textile materials. This concept was initially introduced in the early 1980s by Dr Sweet who was working in Stanford University, United States. With the help of a pressure wave pattern, he could achieve a uniform size and distance among drops (Javorsek and Javorsek 2011). This contributed to a rapid development of inkjet printing. Furthermore, in recent years, inkjet printers have become very popular in the field of printing on textile substrates. A typical sequence of digital textile printing is given in Fig. 6.

The following are the sustainable advantages of digital printing technology.

- It is a clean technology because the inks are utilized to a high degree and thereby minimise water and energy consumption in the aftertreatment.
- Unlimited colour sampling, as well as very good fastness can be obtained.
- Digital printers can switch design without stopping the machine. This is a major advantage when compared to conventional printing where each design requires the making of templates, turning off the machine, adjustment of pattern, and sample printing.
- Reproducibility of designs is very good.
- There is no limitation of size of the repeats of designs.

Fig. 6 Digital printing process of textiles (Chavan 1996)

However, there are some limitations associated with the digital printing of textiles for complete adoption of this technology.

- Printing speed is comparatively slower than conventional rotary screen printing.
- Specific inks with extraordinary quality are to be used which increases the cost of printing.
- Specialized pretreatment of textiles is essential to get very good printed material.

Kan et al. (2011) reported that the atmospheric pressure plasma pretreatment could increase the colour yield of the digital inkjet printed cotton fabrics significantly. In addition, other properties such as colour fastness to crocking, colour fastness to laundering, outline sharpness, and antibacterial properties were also improved when compared with those of the control cotton fabric printed without APP pre-treatment. They found these effects are durable after several washings.

6 Chemical Substitution for Source Reduction

Source reduction is a very useful technique in sustainable processing of textile materials as it will reduce the pollution load in the final effluent and result in a product with lower environmental impact (Ozturk et al. 2009). As per the Okopol Institute, Germany, chemical substitution is defined as 'the replacement or reduction of hazardous substances in products and processes by less hazardous or non-hazardous substances, whilst achieving an equivalent functionality via technological or organizational measures.' Eco-friendly alternatives for chemicals considered as non eco-friendly are summarized in Table 7.

7 Developments in Machinery

Developments in textile machinery for sustainable processing mostly focus on the reduction of water usage in the dyeing process. Currently ultra-low liquor ratio (ULLR) dyeing equipment is available on the market for the processing of textile materials using batch methods. Some of the developments have taken place in reducing energy usage in the processing by combining different stages of processing. One such development in dyeing machinery is the Econtrol® process. Monforts in collaboration with Dystar introduced the Econtrol dyeing process in ITMA in 1995. This technique differs from other reactive dyeing methods in terms of the fixation process where drying and fixation take place in the same step. According to Montfort, the E-control climate inside the fixation chamber ensures a perfect dyeing result during the drying process. By using this process cotton, viscose, tencel, and linen can be dyed. It is claimed that this process uses less energy, water, and chemicals than conventional processes (Ali et al. 2012). The German company M/S Then has introduced a new dyeing machine Airflow Synergy/G2®

Table 7 Chemical substitution for non eco-friendly chemicals

S. no	Purpose	Chemical	Alternative
1	Sizing	Starch	Water soluble polyvinyl alcohol
2	Desizing	Hydrochloric acid	Amylases
3	Scouring of cotton	Sodium hydroxide	Pectinases
4	Bleaching	Hypochlorites	Hydrogen peroxide
5	Oxidation of vat and Sulphur dyes	Potassium dichromate	Hydrogen peroxide, sodium perborate
6	Thickener	Kerosene	Water based polyacrylate copolymers
7	Hydrotropic agent	Urea	Dicynamide (partially)
8	Water repellents	C8 fluorocarbons	C6 fluorocarbons
9	Crease recovery chemicals	Formaldehyde based resin	Polycarboxylic acids
10	Wetting agents and detergents	Alkyl phenol ethoxylates	Fatty alcohol phenol ethoxylates
11	Neutralization agent	Acetic acid	Formic acid
12	Peroxide killer	Sodium thiosulphate	Catalases
13	Mercerization	Sodium hydroxide	Liquid ammonia
14	Reducing agents	Sodium sulphide	Glucose, acetyl acetone, thiourea dioxide
15	Dyeing	Powder form of sulphur dyes	Prereduced dyes
16	Flame retardant	Bromated diphenylethers	Combination of inorganic salts and phosphonates
17	Shrink proofing	Chlorination	Plasma treatment

to get better reduction of water and energy during wet processing. (Nair 2011). M/S Thies GmbH & Co. KG, Germany, introduced the iMaster H2O® dyeing machine for the piece dyeing process. M/S Thies claims that this type of machinery will be helpful for the dyer for better adaptability and flexibility. Globally many such machinery suppliers are available who can supply batchwise processing equipment which will require less energy and water.

8 Role of Eco Standards and Labelling in Promoting Sustainable Processing

Today environmental issues such as global warming and their possible effect on mankind are much in focus. As a result many enlightened customers especially in Europe and other developed countries would like to buy environment friendly and sustainable products even at a higher price. Many eco-labels have been introduced to cater to this class of customers by ensuring the quality and performance of products as well as their safety to human health and the environment by certifying both the products and the manufacturing process. Oekotex and the EU official label for green textiles- EU Flower are two such popular eco-labels. The Global Organic Textile standard (GOTS) whose latest Version 4.0 was introduced in March 2014 aims to set requirements to ensure the organic status of textiles from the harvesting of raw materials through environmentally and socially responsible manufacturing up to labelling to provide credible assurance to consumers. This standard takes into consideration that industrial scale production of textiles is not possible without the use of chemicals. Hence at each manufacturing stage it provides a list of materials which are safe and allowed and which are not allowed to be used. Although allowing the use of a material, minimum impact on the environment, minimum hazard, and toxicity are the criteria. It is in this context that natural dyes and other chemicals from endangered plant species are not permitted but synthetic dyes which can meet toxicity and hazardous substances criteria and do not contain restricted or prohibited substances are allowed to be used. It sets the compliance requirements for the entire facility in respect of environmental management including wastewater treatment and management and also prescribes minimum social criteria.

Eco-labels have taken the initiative to ban or restrict the use of substances which are harmful to human health or pose a risk to the environment. For example, the 'unnecessary use' of triclosan in textiles has been banned by a number of leading brands as well as governments in Europe due to health and ecological issues (Gao and Cranston 2008). Almost all the eco-labels have specified the norms for the maximum release of formaldehyde from the finished textile products. Its release is restricted to 15–20 ppm for babywear and 300 ppm for the textiles which do not have direct contact with human skin. Setting of limits for the harmful substances such as heavy metals, banned carcinogenic amines, brominated and chlorinated flame retardants, alkyl phenol ethoxylates, formaldehyde, and so on in the final product and putting emphasis on the sustainable manufacturing process including the minimum social requirements for certification by these labels thus guarantees a sustainable and quality product to the consumer and contributes to making the textile manufacturing process sustainable. Eco-labels are thus useful for consumers to identify merchandise that has a minimum ecological impact in their life cycle and the number of consumers recognizing eco-labels is increasing and growing.

9 Social Responsibility and Future Outlook in Textile Processing

Social responsibility is one of the main constituents of sustainable development. This is applicable to the highly complex textile value chain also where more players are directly and indirectly involved in it. Figure 7 shows the key stakeholders

Fig. 7 Key stakeholders in textile wet processing value chain

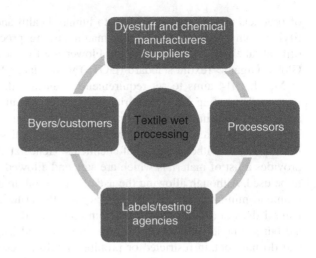

in the textile processing value chain. Sustainable commodity systems will require participation of everybody throughout the chain (Muller 2010). If we consider the wet chemical processing, handling of water and chemicals and their discharge to the environment is a vital parameter in terms of social responsibility which has already been discussed in the chapter. Each stakeholder has his own responsibility in the textile value chain to implement sustainable processing. Dyestuff manufacturers should aim to supply the dyes which are eco-friendly, that is, have high substantivity with low or no requirement of electrolytes. This will result in minimum colour content and low TDS of the residual dye bath after completion of the dyeing process. Intermediates used for the synthesis of dyes should be free from the banned amines or toxic substances.

Textile processers should select eco-friendly dyes, chemicals, and auxiliaries and use such processes and machinery which minimize effluent generation. Suitable mechanism for effluent treatment/recycling of process water and chemicals should also be installed to ensure the minimal effluent discharge and its compliance with the criteria set by regulatory bodies to ensure the availability of community resources such as water for diversified utilisation for the benefit of humanity. Similarly they may install heat exchangers and select the processes with lowest energy consumption to minimise costs and consumption of fossil fuels. They can go for environmental management system standards such as ISO 14000 which will help them to manage environmental related issues in better way. It is also the responsibility of manufacturers to ensure a proper working environment for the workforce. Social Accountability 8000 (SA8000) is an international standard which guarantees the basic rights of workers involved in the production processes. This will improve the product quality and brand reputation.

The main bottleneck in the implementation of sustainable technologies in textile wet processing is their high cost. Buyers' contribution to sustainable development in textile processing is of paramount importance here. If they are ready to purchase the material at the enhanced costs due to the use of costlier sustainable processing technologies, the main barrier to their implementation would be eliminated. Eco-labels help consumers in identifying merchandise that has a minimum ecological impact in their life cycle. The number of consumers recognizing eco-labels is increasing and growing. But due to the multiplicity of eco-labels, they are sometimes confused in the selection of eco-label. Limits of different restricted chemicals in the different labels are also not uniform. Some uniformity and rationalisation of the restricted chemical limits would be helpful in better adoption of eco-labels and provide impetus to sustainable textile wet processing.

10 Conclusion

Chemical processing is the most significant process in the textile value chain. However, it is being criticized for its high energy and water consumption and generation of effluent load. Implementation of stringent government laws related to the environment, introduction of eco-label standards, and competitive market conditions for the past few decades have forced the textile industry to look for new eco alternatives in the form of natural biodegradable and less persistent substances. This scenario has also increased awareness in the industry about newer technologies such as plasma, laser, digital printing, and supercritical carbon dioxide, among others. These new technologies can be considered as sustainable alternatives to conventional technologies because these processes not only reduce water and energy consumption but also significantly reduce the effluent load. Most of these technologies are successful at laboratory and pilot scale; however, issues such as cost economics, nonsuitability for all textile fibres, and so on needs to be resolved for their adoption into mainstream textile processing. It is the duty of each and every stakeholder to promote and adopt sustainable textile processing technologies to ensure a clean environment for future generations.

References

Acxys. http://www.acxys.com/textile.html. Accessed 28 Aug 2014

Agrawal PB, Nierstrasz VA, Warmoeskerken MMCG (2008) Role of mechanical action in low-temperature cotton scouring with F. solani pisi cutinase and pectate lyase. Enzyme Microb Technol 42(6):473–482

Achwal WB (2004) Anti-smell finishes for textiles. Colourage 51(3):33

Ahmed NSE (2005) The use of sodium edate in the dyeing of cotton with reactive dyes. Dyes Pigm 65(3):221–225

Airoli, Plasma (Brochure). http://www.arioli.biz/images/brochure/catalogo_plasma.pdf. Accessed 28 Aug 2014

Anon (2013) APEOs and NPEOs in textiles in O ecotextiles. https://oecotextiles.wordpress.com/2013/01/24/apeos-and-npeos-in-textiles-2/. Accessed 12 Aug 2014

Ali SS, Khatri Z, Brohi KM (2012) Econtrol dyeing process: an ecological and economical approach In: Aslam UM, Khanji H (eds) Energy, environment and sustainable development. Springer, Vienna, pp 291–297

Amorim AM, Gasques MD, Andreaus J, Scharf M (2002) The application of catalase for the elimination of hydrogen peroxide residues after bleaching of cotton fabrics. Anais da Academia Brasileira de Ciências 74(3):433–436

Apjet. http://www.apjet.com/. Accessed 28 Aug 2014

Athalye A (2012) Carbon footprint in textile processing. Colourage 59(12):45–47

Bach E, Cleve E, Schollmeyer E (2002) Past, present and future of supercritical fluid dyeing technology—an overview. Rev Prog Color Relat Top 32(1):88–102

Banchero M (2013) Supercritical fluid dyeing of synthetic and natural textiles—a review. Color Technol 129(1):2–17

Basak S, Saxena S, Chattopadhyay SK, Narkar R, Mahangade R (2015) Banana pseudostem sap: a waste plant resource for making thermally stable cellulosic substrate. J Ind Text. 1528083715591580

Battan B, Dhiman SS, Ahlawat S, Mahajan R, Sharma J (2012) Application of thermostablexylanase of Bacillus pumilus in textile processing. Indian J Microbiol 52(2):222–229

Beck KR, Lynn GM (1997) Extraction of cotton impurities: supercritical CO_2 vs Soxhlet/TCE. Text Chem Color 29(8):70–88

Biradar YS, Jagatap S, Khandelwal KR, Singhania SS (2008) Exploring of antimicrobial activity of Triphala Mashian Ayurvedic formulation. eCAM 5(1):107–113

Blackburn RS, Burkinshaw SM (2003) Treatment of cotton with cationic, nucleophilic polymers to enable reactive dyeing at neutral pH without electrolyte addition. J Appl Polym Sci 89:1026–1031

Börnick H, Schmidt TC (2006) Amines. In: Reemtsma T, Jekel M (ed) Organic pollutants in the water cycle. In: Properties, occurrence, analysis and environmental relevance of polar compounds. Wiley-VCH Verlag GmbH & Co, Germany, pp 181–208

Burkinshaw SM, Lei XP, Lewis DM (1989) Modification of cotton to improve its dyeability. Part 1-pretreating cotton with reactive polyamide-epichloro- hydrin resin. J Soc Dyers Colour 105:391–398

Burkinshaw SM, Mignanelli M, Froehling PE, Bride MJ (2000) The use of dendrimers to modify the dyeing behavior of reactive dyes on cotton. Dyes Pigm 2000(47):259–267

Cai Z, Qui Y, Zhang C, Hwang YJ, McCord M (2003) Effect of atmospheric plasma treatment on desizing of PVA on cotton. Text Res J 73(8):670–674

Chavan RB (1996) Technological revolutions in textile printing. Indian J Fibre Text Res 21(3):50–56

Chuang TH, Wu PL (2007) Cytotoxic 5-alkylresorcinol metabolites from the leaves of Grevillea robusta. J Nat Prod 70(2):319–323

Cookson PG, Brady PR, Fincher KW, Duffield PA, Smith SM, Reincke K, Schreiber J (1995) The Basolan AS process: a new concept in wool dyeing. J Soc Dyers Colour 111(7–8):228–236

Cortez J, Anghieri A, Bonner PL, Griffin M, Freddi G (2007) Transglutaminase mediated grafting of silk proteins onto wool fabrics leading to improved physical and mechanical properties. Enzyme Microb Technol 40(7):1698–1704

Dascalu T, Acosta-Ortiz SE, Ortiz-Morales M (2000) Removal of indigo color by laser beam-denim interaction. Opt Laser Eng 34:179–189

Darnerud PO, Eriksen GS, Jóhannesson T, Larsen PB, Viluksela M (2001) Polybrominated diphenyl ethers: occurrence, dietary exposure, and toxicology. Environ Health Perspect 109(1):49–68

David JM, Zachary RD, Robert JS, Erik H, Jeong-Hoon K, Jung-Gu K, Seong HK (2013) Atmospheric rf plasma deposition of superhydrophobic coatings using tetramethylsilane precursor. Surf Coat Technol 234(15):14–20

Desai JD, Banat IM (1997) Microbial production of surfactants and their commercial potential. Microbiol Mol Biol Rev 61(1):47–64

Dev VG, Venugopal J, Sudha S, Deepika G, Ramakrishna S (2009) Dyeing and antimicrobial characteristics of chitosan treated wool fabrics with henna dye. Carbohydr Polym 75(4):646–650

Diener (2006) Plasma Surface technology. http://www.plasma-us.com/files/diener_web_en.pdf. Accessed 28 Aug 2014

Dow corning corporation (2007) Dow corning plasma solutions' application note. http://www.dowcorning.com/content/publishedlit/01-3137-01.pdf. Accessed 28 Aug 2014

Dyecoo (2010) Dyecoo waterless dyeing http://www.dyecoo.com/pdfs/colourist.pdf Accessed 28 Aug 2015

El-Bendary MA, Abo El-Ola SM, Moharam ME (2012) Enzymatic surface hydrolysis of polyamide fabric by protease enzyme and its production. Indian J Fibre Text Res 37(3):273

El-Sayed II, Kantouch A, Heine E, Höcker H (2001) Developing a zero-AOX shrink-resist process for wool. Part 1: preliminary results. Color Technol 117(4):234–238

Eren HA, Anis P, Davulcu A (2009) Enzymatic one-bath desizing—bleaching—dyeing process for cotton fabrics. Text Res J 79(12):1091–1098

Europlasma (2013) Press release Europlasma launches PFOA- and PFOS-free nanocoatings for techincal textiles under brand name nanofics http://www.europlasma.be/uploads/content/files/PressRelease20130611Techtextil.pdf. Accessed 28 Aug 2014

Fatma AM, El-Alfy EA (2013) Improving dyebility of cotton fabric via grafting with DimethylaminoEthylmethacrylate. J Appl Sci Res 9(1):178–183

Gao Y, Cranston R (2008) Recent advances in antimicrobial treatments of textiles. Text Res J 78(1):60–72. doi:10.1177/0040517507082332

Gebert B, Saus W, Knittel D, Buschmann HJ, Schollmeyer E (1994) Dyeing natural fibers with disperse dyes in supercritical carbon dioxide. Text Res J 64(7):371–374

Gentile DB (2009) A thesis titled Reduced Salt Usage in Dyeing of 100 % Cotton Fabric, School of Fashion & Textiles College of Design & Social Context, RMIT University, July 2009

Guan Y, Zheng Qing-kang, Mao Ya-hong, Gui Ming-sheng, Hong-bin Fu (2007) Application of polycarboxylic acid sodium salt in the dyeing of cotton fabric with reactive dyes. J Appl Polym Sci 105(2):726–732

Han S, Yang Y (2005) Antimicrobial activity of wool fabric treated with curcumin. Dyes Pigm 64:157–161

Han X, Shen T, Lou H (2007) Dietary phenols and their biological significance. Int J Mol Sci 8:950–988

Hartwig H (2002) Plasma treatment of textile fibers. Pure Appl Chem. 74(3):423–427

Hasanbeigi A (2010) Energy-efficiency improvement opportunities for the textile Industry. http://www.energystar.gov/sites/default/files/buildings/tools/EE_Guidebook_for_Textile_industry.pdf. Accessed 12 Aug 2014

Hashem MM (2007) An approach towards a single pretreatment recipe for different types of cotton. Fibres Text Eastern Europe 15(261):85–92

Hebeish A, Ramadan M, Hashem M, Shaker N, Abdel-Hady M (2009) New development for combined bioscouring and bleaching of cotton-based fabrics. Res J Text Apparel 17(1):94–103

Heumann S, Eberl A, Pobeheim H, Liebminger S, Fischer-Colbrie G, Almansa E, Gübitz GM (2006) New model substrates for enzymes hydrolysing polyethyleneterephthalate and polyamide fibres. J Biochem Biophys Methods 69(1):89–99

Hewson MJC (1998) Success with energy management. In: Horrocks AR (ed) Proceedings of the conference, Ecotextile'98, Bolton, Woodhead Publishing, pp 33–34

Hsieh SH, Huang ZK, Huang ZZ, Tseng ZS (2004) Antimicrobial and physical properties of woolen fabrics cured with citric acid and chitosan. J Appl Polym Sci 94:1999–2007

Huang W, Leonas KK (2000) Evaluating a one-bath process for imparting antimicrobial activity and repellency to nonwoven surgical gown fabrics. Text Res J 70(9):774–782

ICAR-CIRCOT, Annual report (2014–15) ICAR-Central institute for research on cotton technology, Mumbai-400019, Maharastra

Javorsek D, Javorsek A (2011) Colour management in digital textile printing. Color Technol 127(4):235–239

Joshi M, Wazed Ali S, Purwar R, Rajendran S (2009) Ecofriendly antimicrobial finishing of textiles using bioactive agents based on natural products. Indian J Fibre Text Res 34(3):295–304

Kalantzi S, Mamma D, Kalogeris E, Kekos D (2010) Improved properties of cotton fabrics treated with lipase and its combination with pectinase. Fibres Text Eastern Europe 18(5):82

Kan CW, Yuen C, Tsoi W (2011) Using atmospheric pressure plasma for enhancing the deposition of printing paste on cotton fabric for digital ink-jet printing. Cellulose 18(3):827–883

Kan CW, Lam CF, Chan CK, Ng SP (2014) Using atmospheric pressure plasma treatment for treating grey cotton fabric. Carbohydr Polym 102(15):167–173

Kan CW, Yuen CWM (2006) Low temperature plasma treatment for wool fabric. Text Res J 76(4):309–314

Kan CW (2015) Washing techniques for denim jeans. In: Paul R (ed) Denim: manufacture, finishing and applications. Elsevier

Karamkar SR (1999) Textile science and technology, Chemical technology in the pre-treatment processes of textiles. Amsterdam, Elsevier

Khatri A, Padhyea R, Whitea M (2012) The use of tri sodium nitrilo triacetate in the pad–steam dyeing of cotton with reactive dyes. Color Technol 129:76–81

Kim J, Kim SY, Choe EK (2006) The beneficial influence of enzymatic scouring on cotton properties. J Nat Fibers 2(4):39–52

Lau C, Anitole K, Hodes C, Lai D, Pfahles-Hutchens A, Seed J (2007) Perfluoro alkyl acids: a review of monitoring and toxicological findings. Toxicol Sci 99(2):366–394

Lee S, Cho JS, Cho G (1999) Antimicrobial and blood repellent finishes for cotton and nonwoven fabrics based on chitosan and fluoropolymers. Text Res J 69:104–113

Lewis DM (2014) Developments in the chemistry of reactive dyes and their application processes. Color Technol 130(6):382–412

Li S, Jinjin D (2007) Improvement of hydrophobic properties of silk and cotton by hexa fluoro propene plasma treatment. Appl Surf Sci 253(11):5051–5055

Liakopoulou-Kyriakides M, Tsatsaroni E, Laderos P, Georgiadou K (1998) Dyeing of cotton and wool fibres with pigments from Crocus Sativus—effect of enzymatic treatment. Dyes Pigm 36(3):215–221

Lim SH, Hudson SM (2004) Application of a fiber-reactive chitosan derivative to cotton fabric as a zero-salt dyeing auxiliary. Color Technol 2004(120):108–113

Long JJ, Wang HW, Lu TQ, Tang RC, Zhu YW (2008) Application of low-pressure plasma pretreatment in silk fabric degumming process. Plasma Chem Plasma Process 28(6):701–713

Mageshwaran V, Walia S, Annapurna K (2012) Isolation and partial characterization of antibacterial lipopeptide produced by *Paenibacillus polymyxa* HKA-15 against phytopathogen *Xanthomonas campestris* pv. phaseoli M-5. World J Microbiol Biotechnol 28(3):909–917

Makovitzki A, Avrakhami D, Shai Y (2006) Ultra short antibacterial and antifungal lipopeptides. Proc Natl Acad Sci 103(43):15997–16002

Man WS, Kan CW, Ng SP (2014) The use of atmospheric pressure plasma treatment on enhancing the pigment application to cotton fabric. Vacuum 99:7–11

Marbek Resource Consultants (2001) Identification and evaluation of best available technologies economically achievable (BATEA) for textile mill effluents. http://www.p2pays.org/ref/41/40651.pdf. Accessed 20 Aug 2015

Matamá T, Carneiro F, Caparrós C, Gübitz GM, Cavaco Paulo A (2007) Using a nitrilase for the surface modification of acrylic fibres. Biotechnol J 2(3):353–360

Mehmood T, Kaynak A, Dai XJ, Kouzani A, Magniez K, de Celis DR, Du Plessis J (2014) Study of oxygen plasma pre-treatment of polyester fabric for improved polypyrrole adhesion. Mater Chem Phys 143(2):668–675

Menezes E, Choudhari M (2011) Pre-treatment of textiles prior to dyeing. In: Hauser P (ed) Textile dyeing InTech. http://www.intechopen.com/books/textile-dyeing/pre-treatment-of-textiles-prior-to-dyeing. Accessed 12 Aug 2014

Min BR, Pinchak WE, Merkel R, Walker S, Tomita G, Anderson RC (2008) Comparative antimicrobial activity of tannin extracts from perennial plants on mastitis pathogens. Sci Res Essay 2:66–73

Miyazaki K, Tabata I, Hori T (2012) Relationship between colour fastness and colour strength of polypropylene fabrics dyed in supercritical carbon dioxide: effect of chemical structure in 1,4-bis(alkylamino)anthraquinone dyestuffs on dyeing performance. Color Technol 128(1):60–67

Mohamed AL, Rafik ME, Moller M (2013) Supercritical carbon dioxide assisted silicon based finishing of cellulosic fabric: a novel approach. Carbohydr Polym 98(1):1095–1107

Montazer M, Maryan AS (2010) Influences of different enzymatic treatment on denim garment. Appl Biochem Biotechnol 160(7):2114–2128

Muthu SS, Li Y, Hu JY, Ze L (2012) Carbon footprint reduction in the textile process chain: Recycling of textile materials. Fibers Polym 13(8):1065–1070

Muller L (2010) Understanding the cotton value chain. http://www.lizmuller.com/uploads/2/0/1/0/20101265/understanding_the_cotton_supply_chain.pdf. Accessed 25 Aug 2015

Nair GP (2011) Methods and machinery for the dyeing. In: Clark M (ed) Handbook of textile and industrial dyeing: principles, processes and types. Wood head publishing, pp. 291–293

Naveed S, Bhatti I, Ali K (2006) Membrane technology and its suitability for treatment of textile waste water. Pakistan. J Res (Science) 17(3):155–164

Ondogan Z, Pamuk O, Ondogan EN, Ozguney A (2005) Improving the appearance of all textile products from clothing to home textile using the laser technology. Opt Laser Technol 37(8):631–637

Ozturk E, Yetis U, Dilek FB, Demirer GN (2009) Chemical substitution study for a wet processing textile mill in Turkey. J Clean Prod 17(2):239–247

Pazarlioğlu NK, Sariişik M, Telefoncu A (2005) Treating denim fabrics with immobilized commercial cellulases. Process Biochem 40(2):767–771

Plasmatreat. http://www.plasmatreat.com/. Accessed 28 Aug 2014

Prabu HG, Sundrarajan M (2002) Effect of the bio-salt trisodium citrate in the dyeing of cotton. Color Technol 118(2002):131–134

Purwar R, Joshi M (2004) Recent Developments in antimicrobial finishing of textiles-a review. AATCC Rev 4:22–26

Ramesh Babu B, Parande AK, Raghu S, Prem Kumar T (2007) Textile technology: cotton textile processing waste generation and effluent treatment. J Cotton Sci 11(3):141–153

Raja ASM, Thilagavathi G (2010) Comparative study on the effect of acid and alkaline protease enzyme treatments on wool for improving handle and shrink resistance. J Text Inst 101(9):823–834

Samanta KK, Jassal M, Agrawal AK (2010) Atmospheric pressure plasma polymerization of 1, 3-butadiene for hydrophobic finishing of textile substrates. J Phys: Conf Ser 208:012098

Saravanan D (2007) UV protection textile materials. AUTEX Res J 7(1):53–62

Saravanan D, Sivasaravanan S, SudharshanPrabhu M, Vasanthi NS, Senthil Raja K, Das A, Ramachandran T (2012) One-step process for desizing and bleaching of cotton fabrics using the combination of amylase and glucose oxidase enzymes. J Appl Polym Sci 123:2445–2450

Savarino P, Montoneri E, Bottigliengo S, Boffa V, Guizzetti T, Perrone DG, Mendichi R (2009) Biosurfactants from urban wastes as auxiliaries for textile dyeing. Ind Eng Chem Res 48(8):3738–3748

Sawada K, Jun JH, Ueda M (2003) Dyeing of natural fibres from perfluoro-polyether reverse micelles in supercritical carbon dioxide. Color Technol 119(6):336–340

Seong HS, Kim JP, Ko SW (1999) Preparing chito-oligosaccharides as antimicrobial agents for cotton. Text Res J 69(7):483–488

Shin JH, Baek YJ (2012) Analysis of polybrominated diphenyl ethers in textiles treated by brominated flame retardants. Text Res J 82(13):1307–1316. doi:10.1177/0040517512439943

Shishoo R (2007) Introduction-Potential application of plasma technology in the textile industry. In: Shishoo R (ed) Plasma technologies for textile. Wood head Publishing, Cambridge

Sigma technologies (2006) Sigma technologies atmospheric plasma treaters for high-speed web applications. http://sigmalabs.squarespace.com/storage/publications-and-resources/SIGMA%20APT%20Brochure.pdf. Accessed 28 Aug 2014

Singh R, Jain A, Panwar S, Gupta D, Khare SK (2005) Antimicrobial activity of some natural dyes. Dyes Pigm 66(2):99–102

Softol. http://www.softal.de/de/loesungen/oberflaechenbehandlung-mit-der-aldyne-technologie/. Accessed 28 Aug 2014

Sun D, Stylios GK (2004) Effect of low temperature plasma treatment on the scouring and dyeing of natural fabrics. Text Res J 74(9):751–756

Tarhan M, Sarıışık M (2009) A comparison among performance characteristics of various denim fading processes. Text Res J 79(4):301–309

Thilagavathi G, Kannaian T (2010) Combined antimicrobial and aroma finishing treatment for cotton, using microencapsulated geranium (Pelargonium graveolensL'Herit, ex Ait.) leaves extract. Indian J Nat Prod Resour 1(3):348–352

Tzanov T, Costa S, Guebitz GM (2001) Dyeing in catalase treated bleaching baths. Color Technol 117(1):1–5

Van der Kraan M (2005) Process and equipment development for textile dyeing in supercritical carbon dioxide. Delft University of Technology, TU Delft

Vitoplasma. http://www.vitoplasma.com/en/plasmazone. Accessed 28 Aug 2014

Wang H, Lewis DM (2002) Chemical modification of cotton to improve fibre dyeability. Color Technol 118:159–168

Wasif AI, Indi IM (2010) Combined scouring and bleaching of cotton using potassium persulpahte. Indian J Fibre Text Res 35:353–357

Wei MA, Shu-fen Z, Zong YJ (2012) Application mechanism and performance of cationic native starch and cationic hydrolyzed starch in salt-free dyeing of reactive dyes. Appl Mech Mater 161:212–216

Wen H, Dai JJ (2007) Dyeing of polylactide fibers in supercritical carbon dioxide. J Appl Polym Sci 105(4):1903–1907

World Commission on Environment and Development (1987) Our common future. Oxford University Press, Oxford

Wu TS, Chen KM (1993) New cationic agents for improving the dyeability of cotton fibers. J Soc Dyers Colour 109:153–157

Yu B, Wang WM, Cai ZS (2014) Application of sodium oxalate in the dyeing of cotton fabric with reactive red 3BS. J Text Inst 105(3):321–326

Zaffalon V (2010) Climate change, carbon mitigation and textiles. Textile World. http://textileworld.com/Articles/2010/July/July_August_issue/Features/Climate_Change_Carbon_Mitigation_In_Textiles.html. Accessed 12 Aug 2014

Zaharia C, Suteu D, Muresan A, Muresan R, Popescu A (2009) Textile wastewater treatment by homogenous oxidation with hydrogen peroxide. Environ Eng Manage J 8(6):1359–1369

Zhang F, Chen Y, Lin H, Lu Y (2007) Synthesis of an amino-terminated hyperbranched polymer and its application in reactive dyeing on cotton as a salt-free dyeing auxiliary. Color Technol 123:351–357

Printed in the United States
By Bookmasters